PENGUIN BOOKS

ALPHA & OMEGA

Charles Seife, a journalist who holds a master's degree in mathematics from Yale University, writes for *Science* magazine and has also written for *New Scientist*, *Scientific American*, *The Economist*, *Wired*, *The Sciences*, and many other publications. His previous book, *Zero*, won the PEN/Martha Albrand Award for first nonfiction book and was named a *New York Times* Notable Book. He lives in Washington, D.C.

A alpha

[CHARLES SEIFE]

Penguin Books

The Search for the Beginning and End of the Universe

omega

PENGUIN BOOKS

Published by the Penguin Group

Penguin Group (USA) Inc., 375 Hudson Street, New York, New York 10014, U.S.A.
Penguin Books Ltd, 80 Strand, London WC2R 0RL, England
Penguin Books Australia Ltd, 250 Camberwell Road, Camberwell, Victoria 3124, Australia
Penguin Books Canada Ltd, 10 Alcorn Avenue, Toronto, Ontario, Canada M4V 3B2
Penguin Books India (P) Ltd, 11 Community Centre,
Panchsheel Park, New Delhi – 110 017, India
Penguin Books (N.Z.) Ltd, Cnr Rosedale and Airborne Roads,
Albany, Auckland, New Zealand
Penguin Books (South Africa) (Pty) Ltd, 24 Sturdee Avenue,
Rosebank, Johannesburg 2196, South Africa

Penguin Books Ltd, Registered Offices:
80 Strand, London WC2R 0RL, England

First published in the United States of America by Viking 2003
Published in Penguin Books 2004

1 3 5 7 9 10 8 6 4 2

Grateful acknowledgment is made for permission to reprint
excerpts from the following copyrighted works:
"Cosmic Gall" from Collected Poems 1953–1993 by John Updike. Copyright © 1993 by
John Updike. Used by permission of Alfred A. Knopf, a division of Random House, Inc.
"The End of the World" from Collected Poems 1917–1982 by Archibald MacLeish.
Copyright © 1985 by The Estate of Archibald MacLeish.
Reprinted by permission of Houghton Mifflin Company. All rights reserved.

Photograph credits
European Southern Observatory: Insert page 1 (all), page 2 (top left)
Space Telescope Science Institute: Page 2 (top right)
Goddard Space Flight Center/National Aeronautics and Space Administration:
Page 2 (bottom)

Drawings by Matt Zimet

THE LIBRARY OF CONGRESS HAS CATALOGED THE HARDCOVER EDITION AS FOLLOWS:
Seife, Charles.
Alpha and omega : the search for the beginning and
end of the universe / Charles Seife.
p. cm.
Includes index.
ISBN 0-670-03179-8 (hc.)
ISBN 0 14 20.0446 4 (pbk.)
1. Cosmology. 2. Astronomy. I. Title.
QB981.S446 2003
523.1 — dc21 2002044853

Printed in the United States of America
Set in Cochin
Designed by Jaye Zimet

Contents

Preface

I am Alpha and Omega, the beginning and the end, the first and the last.

—REVELATION 22:13

Ten billion light-years away, Nature screams. In a fraction of a second, a star explodes with more energy than ten billion billion billion hydrogen bombs. For a few weeks, the funeral pyre of the dying sun blazes and outshines the countless stars of its galaxy. When a star dies as a supernova, it is visible halfway across the universe.

The light from that supernova travels for ten billion years, attenuated and stretched along the way. By the time the light reaches Earth, it is far too dim to be spotted by the naked eye, but telescopes can see the supernova as a dim blotch in the sky. It is a message from the ends of the cosmos—a message whose receipt on Earth heralds the beginning of a revolution.

This revolution began in the late 1990s, when two teams of scientists began to decode the death throes of dying stars. Their observations showed that the universe was suffused with a mysterious "dark energy," an invisible substance that stretches the very fabric of space and time. The discovery of dark energy baffled and delighted astronomers, who scrambled to confirm the observations and understand the enigma. What's more, the stellar death rattles held the secret to the universe's death—scientists merely had to decrypt the message from the dying stars and they would understand how the cosmos would end.

That message has now been deciphered. On June 25, 2001, *Time* magazine devoted its cover to the end of the universe. "Peering deep into space and time, scientists have just solved the biggest mystery in the cosmos," it exclaimed. This is no overstatement. Cosmologists now know how the universe will end, and a new set of experiments, whose results have begun to trickle out, is removing the veil over the big bang, showing us how it began.

The revolution is being fought on many fronts, by astronomers, cosmologists, and physicists, high atop the Chilean mountains, deep underneath the Canadian soil, stranded in the middle of the Antarctic wasteland, and all across the globe. *Alpha and Omega* is the story of galaxy hunters and the microwave eavesdroppers, gravity theorists and particle physicists, quantum theorists and atom smashers, all of whom are on the brink of major discoveries. Each of their stories, taken alone, would be noteworthy. Together, they add up to a renaissance—a major shift in our understanding of the universe. This shift is happening right now, and it is far from finished.

Alpha and Omega is the story of the most exciting scientific discoveries in decades and the people behind them. It is also a guide to understanding the headlines that are erupting in *Time,* in the *New York Times,* in *Science,* and in newspapers and magazines all across the planet. This revolution in cosmology will be front-page news again and again over the next few years. Indeed, it will be one of the most important scientific stories of the twenty-first century. When it is over, we will have seen the moment of creation, and we will see the face of our own destruction.

Chapter 1
The First Cosmology

[THE GOLDEN AGE OF THE GODS]

*Then All-father took Night and her son, Day, and gave
them two horses and two chariots and put them up in the
sky, so that they should ride round the world every
twenty-four hours. Night rides first on a horse called
Hrimfaxi, and every morning he bedews the earth with
the foam from his bit. Day's horse is called Skinfaxi, and
the whole earth and sky are illumined by his mane.*

— SNORRI STURLUSON, *THE PROSE EDDA*

Perhaps it happened on a midwinter's night thirty thousand
years ago. A tribe of cavemen huddled close to the embers
of a dying flame. A single hairy face gazed upward, bewildered.
Against the innumerable, immutable pinpricks of light in the
heavens, a star had moved. A human looked into the cosmos
and saw the trail of a wandering god.

Even before the dawn of civilization, people gazed sky-
ward and wondered. Who created the stars in the sky? How

was the universe born? Will it end? If so, how? These are the most ancient questions of humanity. Yet, for millennia upon millennia, the only way to answer these mysteries was through mythology. Even today, the remnants of that mythology can be seen in the heavens. The tiny lights that meander slowly through the sky, better known as planets, bear the names of gods. Red Mars is gorged with the blood of conquest; bright Venus glitters in the morning with the allure of the goddess of love. Each civilization invoked its own gods to explain the creation of the universe, the existence of stars in the night sky, and occasionally the ultimate destruction of the cosmos.

Three revolutions separate modern cosmologists from the shamans and storytellers of the age of mythology. The first, which took place in the 1500s, was the most dangerous. Its enemies tried to stifle it with all the weapons in their arsenal: accusations of heresy and witchcraft. The second revolution, which began in the 1920s, was the most unsettling; the comforting concept of a clockwork universe was shattered, and humanity was suddenly alone in a vast, empty cosmos. For the first time, scientists saw evidence of the act of creation. These two revolutions take us to the present day, where we are in the midst of a third revolution, a revolution that is finally answering the eternal questions, revealing our origins and our ultimate fate.

If you look upward on a sunny day and squint your eyes just right, you can imagine the vault of the heavens as an immaculate blue dome, arching high above the wispy clouds that float slowly across the sky. To ancient peoples, the dome of the sky was a real object; the Earth was enclosed by a beautiful sphere that shone blue in the daytime as the sun slowly traveled from east to west. In the evening, tiny, flickering points of light mocked the humans far below, and a faint shimmering ribbon stretched across the giant ball surrounding the Earth.

Who fashioned that sphere? Each culture had a different answer; every people had a story of creation, which told of how the gods came to be and how they created the universe. The Norse people, not surprisingly, thought that the universe was born from ice. As the frost encountered an enormous fire, it thawed and formed a giant named Ymir. Odin, chief of the gods, and his brothers slew Ymir and used his skull as the dome of heaven. They then fashioned the Earth from Ymir's flesh, the oceans from his blood, and the clouds from his brains. They set the planets in the sky and made the glowing chariots of the sun and moon chase each other in the vault of the heavens —each eternally pursued by a wolf.[1] The Pawnee Indians of central North America saw corn as the mother of all things; Mother Corn gave life to humanity, which emerged from the ground like the crops that the Pawnee depended on. Some cultures thought the universe began as a vast ocean; others, as a shapeless chaos. There are dozens and dozens of vastly different tales of the creation of the universe, but most of them focus on the same events: the birth of the gods; the creation of the heavens, Earth, and stars; and the fashioning of man and woman. These elements are the foundation of any religion, as they answer the fundamental questions that humans have been asking since the dawn of time. Before the scientific revolution gave humanity another tool with which to examine the universe, people could only explore its history and nature by listening to the stories of shamans and the musings of philosophers. Religion and philosophy formed the cosmologies of the ancients.

Two of these numerous cosmologies dominated the Western world, from before the ascent of Rome until the time of William Shakespeare. Even though these two traditions are mutually contradictory, they fused, and fashioned a story of the universe that was almost unassailable until the advent of the scientific method. The combination of an Eastern, Semitic

1. Unfortunately for the sun and moon, the wolves catch up in the end.

cosmology, encoded by the Bible, and a Western, Greco-Roman one, became a solid structure that stood for more than a millennium. It took a cosmological revolution to tear the edifice down.

The word *cosmos* is the Greek word for "order," and the cosmos—the universe as a whole—was the only order to be found in the chaos of Greek mythology. The sun traveled across the sky each day, guided by Helios, the solar charioteer.[2] The moon waxed and waned each month, growing pregnant and barren in turn. And in the night sky, the stars remained fixed, except for five wanderers—the planets—that moved across the unchanging backdrop of the heavens.[3] Even today, we know the planets by their Olympian names: Mercury, Venus, Mars, Jupiter, and Saturn are the Roman names of the Greek gods Hermes, Aphrodite, Ares, Zeus, and Cronus. The Greeks saw order in the clockwork motions of the heavenly bodies, and from early on in their civilization they began to work out the details of that clockwork. In 585 BC, the Greek mathematician Thales was the first to predict the coming of a solar eclipse. According to Herodotus, two warring peoples, the Medes and the Lydians, were astonished to see the day turn into night and decided that it would be a good time to put down their weapons.

By trying to understand how the heavens worked, Thales became the first starry-eyed cosmologist—to the amusement of his neighbors. "While he was studying the stars and looking upward, he fell into a pit, and a neat, witty Thracian servant girl jeered at him," Socrates reportedly said, several centuries later. But Thales put all his concentration and observation to good use. He created an entire cosmos from the sheer power of his mind.

2. One legend tells of a single, disastrous aberration in the sun's daily course, when the son of Helios took the reins of the chariot. The son, Phaëthon, died due to his hubris and poor horsemanship.
3. The Greek word *planetos* means "wanderer."

Perhaps because the Greek stories of creation were fragmentary and contradictory, Thales ignored them when building his cosmology. Though he believed that gods were everywhere in the universe, Thales took the act of creation out of the gods' hands. In Thales' universe, water was the source of all things; earth floated upon the water like a cork. Not everyone agreed with Thales that water was the primordial material from which the universe was made. Others, like Anaxagoras and Diogenes, argued that air came before water. (After all, water destroys fire, so water could hardly have given birth to fire.) Yet others argued that fire was prime. Empedocles, who lived at around 450 BC, refused to pick a single primal essence and instead argued that earth, air, fire, and water were the four elements. In different combinations, he declared, these four essences made up everything in the universe.

The philosophers also argued about the nature of the heavenly clockwork. They looked to the heavens and tried to figure out the order of the cosmos, and Earth's place within that order. They began by describing the Earth itself. Pythagoras, an eccentric philosopher who is best known for his theorem about right triangles, argued that the planets, including Earth, revolved around a central fire. Others argued that the Earth was flat, and still others that it was spherical, but at the center of the universe. By the fourth century BC, Aristotle became the philosopher who mattered. Born in Macedonia, and tutored by Socrates' student Plato, Aristotle, in turn, became the teacher of Alexander of Macedon—better known as Alexander the Great. And just as surely as Alexander conquered the West, so too did Aristotle's philosophy.

Aristotle's cosmos was exquisitely orderly. Everything had its place in the universe. Empedocles' four elements had their natural positions; earth, the heaviest element, sank to the center of the universe, so the Earth, quite naturally, must be at the very center of the cosmos. Water was slightly lighter, so it floated above earth, but below air and fire, which were ᵤ..ter still. Aristotle added a fifth element—literally, the quintessence—that was purest of all. Earthly things were made of earth, air, fire, and water; the quintessence was only found in the heavens. To Aristotle, the pure, unchanging heavens were made of stuff entirely different from the ever mutable, but motionless, Earth at the center of the universe. The moon, sun, and planets each revolved around the Earth in perfect, crystalline spheres, never ceasing in their motion, and filling the heavens with celestial harmony: the music of the spheres.

This cosmology was based upon pure logic. Aristotle made certain basic assumptions—that the universe had to be finite, that everything had a natural place, that circles and spheres were the most perfect geometric shapes—and deduced what he thought was the natural order of the cosmos. Aristotle's mentor, Plato, mocked the "light-minded men" who, "being students of the worlds above, suppose in their simplicity that the most solid proofs about such matters are obtained by the sense of sight," and Aristotle agreed. Observation was for fools.

Aristotle's cosmos was light on theology. It only required the existence of a "prime mover" to set the celestial spheres in motion—it did not specify the nature of that divine power. This, in part, is what gave Aristotle's cosmos such longevity even after an entirely different culture became the foundation for Western religion.

"In the beginning God created the heaven and the earth. And the earth was without form, and void; and darkness was upon the face of the deep. And the Spirit of God moved upon the face of the waters." The beginning of Genesis is the basis for Jewish—and, later, Christian—cosmology. Its roots lie in the hazy past of the first civilization, in the Fertile Crescent. Thousands of years later after the Hebrew Bible was set down in writing, Christ took this ancient tradition and bent it into a new form.

Unlike the Greek cosmology, which could easily accomo-
date a pantheon of petty, squabbling gods, the Jewish cosmology tells of an omnipotent, omniscient God who creates
the heaven and Earth out of nothing. He alone fashions the
vault of heavens and the Earth below; he alone set the sun,
moon, and stars in their places in the sky. His act of creation
took six days, but the universe, complete with the heavenly
bodies, was finished by the fourth. God created man on the
sixth day—the culmination of his efforts.[4] The hierarchy is
clear. Genesis sets it out quite neatly. God is above all, and
then comes man, which God created in his image. Then comes
woman. Then the beasts of the field, the fowl of the air, the fish
of the sea, herbs and plants, and then the Earth itself. Man has
dominion over all; everything else in the universe is meant to
serve him. The sun and moon were meant to divide the night
from the day for man's benefit; along with the innumerable
stars, they were fashioned to provide him with light. Man is
the center of the universe, both literally and figuratively.

When Rome conquered Greece, it absorbed Greek philosophy and culture—and its cosmology—and as the Roman
Republic and Empire spread across the known world, so too
did Aristotle's picture of the universe. But Rome, in turn,
would be conquered by Christianity, a religion that branched
off of Judaism. At the end of the first century AD, Christianity
was a small sect. Less than three centuries later, the emperor
Constantine, ruler of the most powerful nation on earth, converted to Christianity. The Greco-Roman and Christian cultures began to merge. Aristotle's ambiguous theology made it
easy for the early Christians to absorb Aristotle, just as Rome
had. (The New Testament was written in Greek, after all, so

---

4. The two Genesis stories of the origin of man and woman are somewhat contradictory. Genesis 1 has both man and woman being created on the sixth day; Genesis 2
starts with Adam and tells of Eve being fashioned from Adam's rib. For this reason,
some Jewish mystics believed that Adam had a wife before Eve, Lilith, who now
wanders the Earth as a demon.

the early church had already absorbed a heavy dose of Greek culture.) Christianity, with Greek philosophical undertones, became the dominant cosmology in the Western world.

The Aristotelian component of Western cosmology had a very firm foundation—it was based upon observation of the natural world. In the second century AD, in Alexandria, the intellectual capital of the ancient world, the mathematician Ptolemy built an intricate, and incredibly complicated, model of the universe based upon Aristotle's cosmology. The Earth was at the center of the universe, and the stars and planets whirled in circular orbits around it. To explain the complicated motions of the planets (such as the occasional backward, or *retrograde,* motion of Mars), Ptolemy proposed that the planets danced in tiny little circles called epicycles as they spun around the Earth.

Ptolemy's clockwork universe worked beautifully. It explained the motions of the planets to fairly high precision, providing seemingly unshakeable support for Aristotle's theory of the cosmos. By building upon Aristotle's geocentric universe, Ptolemy had fashioned a powerful cosmology, one that had predictive power. Its ability to describe the motion of the planets, along with its "prime mover" that seemed to describe the Christian God admirably well, made the Aristotelian-Ptolemaic universe unassailable until Elizabethan times.

Aristotelian-Ptolemaic cosmology was embraced by the church, even though it sometimes contradicted the Bible. For instance, Psalm 148 exclaims, "Praise him, ye heavens of heavens, and ye waters that be above the heavens." Though having a water above the heavens seemed to explain both the blueness of the sky and the source of rain, this was forbidden in the Aristotelian universe. Water is a heavy element, so it did not belong above the heavens; it was only allowed to exist in the earthly sphere.

Though the church struggled internally with the contradictions between Aristotle and the Bible, it eventually used

Aristotelian cosmology as the basis for its own theology. To attack Aristotle became tantamount to attacking the truths handed down by the pope himself. And when a revolution toppled Aristotle, the church found itself on the losing side. It has never recovered.

Chapter 2
The First Cosmological Revolution
[THE COPERNICAN THEORY]

The indispensable catalyst is the word, the explanatory idea. More than petards or stilettos, therefore, words — uncontrolled words, circulating freely, underground, rebelliously, not gotten up in dress uniforms, uncertified — frighten tyrants.

— RYSZARD KAPUSCINSKI, *SHAH OF SHAHS*

High up on a wall in the Vatican, there is a tiny yellow portrait, a relic of a four-hundred-year-old battle. Surrounded by flowers and laurels, and crested by the two keys to heaven, a bemused-looking bearded man gazes to his left. He would be unrecognizable but for a handy Latin inscription: "Galileus." Galileo Galilei, the most famous scientist of his day, was condemned to perpetual imprisonment by the Roman Inquisition. Now that he's back in favor with the church, his picture is adorned with the trappings and symbols of the pontiff. A few yards away, another bearded man on the wall stares to his right. His three-pointed hat reveals that he is a cardinal, a prince of the church. "Bellarminus," Cardinal Robert Bellarmine, the chief of the Roman Inquisition, was the man who first tried to subdue Galileo. His portrait is also surrounded by laurels, and he too is adorned by the keys to heaven. Galileo and Bellarmine, adversaries in life, are both honored by the church, and their portraits adorn the same wall of the Vatican. Yet the two still look in opposite directions.

Four hundred years after the opening shots of the first cosmological revolution, the Roman Catholic Church is still struggling to come to terms with its past. When the discipline of science was born, the church tried to smash the scientists who trod carelessly across Christian theology. Unfortunately for some scientists, it was hard not to stray into forbidden territory, especially since the first major achievement of modern science was to shatter the ancient Aristotelian cosmos—to smash the cozy universe, as self-contained as a nutshell—into a thousand shards. For the first time, science provided genuine insight into the nature of the universe. A new breed of philosopher started telling a tale of how the cosmos was put together, contradicting the Aristotelian-Ptolemaic cosmology. The church, with its foundation perched upon that ancient nutshell, struck back.

The scientific cosmology eventually defeated the Aristotelian one but this would not provide any consolation for Galileo, or for the other victims of that struggle. Centuries after the clash of theology with science, the church still suffers from losing the first cosmological revolution.

The church had a love-hate relationship with Aristotle and Ptolemy, the ancient Greek architects of Western cosmology. The Greek cosmos made a great deal of sense to the medieval mind; the stars and planets had their own natural places. So did the elements that made up all the matter in the cosmos. Heavy earth sinks to the center of the universe, forming the ground we walk on. Water, which is lighter, sits on top of the earth, forming the oceans and rivers. Air, lighter still, forms the atmosphere we breathe. Fire is the lightest element—after all, flames try to leap into the sky. The sun, moon, planets, and stars, made from some light, fiery substance, inhabit the heavens, revolving in crystal spheres about our planet. What made the cosmology even more attractive to the church was that Aristotle's universe required a prime mover. The Greek cosmology was inherently a proof of a divine existence.

Few philosophers or theologians in the West questioned the idea that the prime mover, the being who set the crystal spheres in motion, was the Christian God. The church embraced Aristotle's ideas, seeing the value of a proof of God's existence. But before long, theologians realized that the Aristotelian cosmology, in fact, contradicted the Bible. The ancient Greek wisdom denied the existence of an omnipotent God—a heretical idea. Cracks in Aristotle's nutshell universe began to appear very early in church history. Augustine of Hippo, the fifth-century scholar and saint, was one of the first to attack the ancient philosophy.

Augustine saw a problem with Aristotle's ideas about the origin of motion, which comes from the prime mover. That idea in itself was not so troublesome to Augustine, but the devil was in the details. Aristotle's prime mover twists the outermost crystal sphere, causing the motion of all things in the universe, be it the eternal spinning of the planets in the heavens or the motion of the flames on a burning piece of flax. Therefore, argued the ancient philosophers, if God decided to stop the motion of the heavens, then all motion on Earth and the very passage of time should cease. Water should abruptly stop flowing over a waterfall, and birds should freeze motionless in midflight. Since God doesn't stop the motions of the heavens very often, this doesn't seem like a theory that would get tested. But, in fact, it did, in a medieval sort of way. God did stop the motion of the heavens once, at least according to the Bible. And biblical history does not match up with Aristotle's predictions.

Chapter 10 of the Book of Joshua tells of a battle between the Israelites and the inhabitants of the land of Canaan: "And the sun stood still, and the moon stayed, until the people had avenged themselves upon their enemies." The men of Israel were happily smiting and slaying, even though the heavenly bodies had stopped in their tracks. This biblical passage directly contradicts Aristotle, whose theory implies that the Israelites should have been as motionless as the sun and moon.

Realizing the contradiction between the Bible and Greek philosophy, Augustine argued that the passage of time was independent of the motion of the heavenly bodies; if the sun and moon stood still in the heavens, a potter's wheel would still whirl around unabated. If Aristotelian philosophy and the Bible clashed, then Aristotle had to give way.

The friction between Aristotle and the Bible was inevitable. The Bible is based on Eastern philosophy, while medieval cosmology was built upon the Western philosophy of Aristotle and his successors. The two cultures had very different views about how the universe works, yet the two were forced into an uneasy marriage within church theology. The inherent contradictions led to centuries of conflict, which reached a peak in the 1200s.

Theologians argued, in the tradition of Augustine, that an omnipotent God can do whatever he wants to do; if he so desires, he can stop the planets and maintain the flow of time. He can create a void or a vacuum, an act absolutely forbidden by Aristotle's philosophy. (This abhorrence of the void forced Aristotelian scholars to conclude, rather absurdly, that all motion had to be circular—moving in a straight line was impossible. Motion in a straight line would create a vacuum behind the moving object. With circular motion, on the other hand, everything simply swapped positions without creating a vacuum.) However, Aristotle's no-vacuum declaration directly contradicts the Bible. Genesis says that the universe was born from void, a concept that Aristotle would have found to be ridiculous. Aristotle's rules were shackles on the hands of a

God who is too powerful to be shackled. Therefore, some clerics concluded, Aristotle must be wrong. In the first half of the thirteenth century, one cardinal banned Aristotle's *Physics* as well as his *Metaphysics*. Shortly thereafter, in 1277, the bishop of Paris, Étienne Tempier, called a council together to refute elements of Aristotelian cosmology, such as "God cannot move the heavens in a straight line, because that would leave behind a vacuum." If God wants to move the heavens in a straight line, argued Tempier, who could stop him? Certainly not Aristotle. The council condemned the "errors" that led adherents of Greek philosophy into heresy.

The pro-Aristotle camp fought back, particularly Thomas Aquinas, a noble-born hermit who found the philosophy of the ancients aesthetically—and theologically—pleasing. He argued for a deeper integration of Aristotelian cosmology within church theology. Aquinas died in 1274, after many years of serious research into such things as how many angels can dance on the head of a pin.[1] Three years later, some of his statements were condemned by Tempier as heresy. But in 1323, Aquinas got a promotion. *Saint* Thomas Aquinas could hardly be a heretic, so Tempier's condemnations fell by the wayside. Nonetheless, the question of Aristotle's role in the church was far from settled. The battle raged back and forth. For a time, even those at the very center of the church were espousing radical anti-Aristotelian ideas.

In the fifteenth century, Nicholas of Cusa, a cardinal, argued that the glittering stars in the sky were like our own sun; perhaps each point of light in the firmament was a distant solar system of its own, complete with alien Earths. Perhaps those alien Earths even had their own moons. This was a direct challenge to Aristotle, to the very idea that every element in the universe has its natural place. In Aristotle's cosmology,

1. Aquinas used Aristotelian philosophy to deduce that two angels can't be in the same place at the same time (the medieval equivalent of the Pauli exclusion principle, which was formulated in 1925). This idea was caricatured by later philosophers as the angels on pins argument.

Earth must be unique, because there can be only one reposi-
tory of earth, the heaviest element. All heavy objects, like
rocks, goats, trees, and people, try to sink to the center of the
universe and are held up only by the elemental earth that
forms the ground beneath our feet. Only things made out of
lighter elements, like air and fire, can float in the sky. So, in
Aristotelian cosmology, the very idea of other Earths is absurd. Any dirt in the sky would instantly crash down on our
heads as it assumed its natural place at the center of the
universe. Cusa, on the other hand, stated that other worlds—
bits of rock and soil—float in the heavens. It was an absurd
theory to any Aristotelian.

Cusa did not stop there. He boldly declared that all of
these alien worlds had inhabitants. There were infinite worlds
in the cosmos, teeming with infinite alien beings. Maybe
those aliens gaze up in the sky at night and look at a point of
light, our Earth, and wonder whether the tiny gleaming pin-
prick could harbor life. If so, how could the Vatican be the
seat of the One True Church? How could alien beings obey
the pope if they had never heard of Rome? Cusa's doctrine
was very dangerous to the church, but it escaped notice, even
after 1543 when a Polish clergyman, Nicolaus Copernicus,
gave scientific support for Cusa's bold statement: the Earth is
not the center of the universe. The church did not realize at
the time that the first cosmological revolution had begun.

Science would never have come into conflict with the church
if the scientific revolution had dealt with a discipline that was
largely devoid of spiritual implications, like botany or chem-
istry. But scientists ventured into cosmology, a very touchy
subject because it was traditionally the territory of theologians and philosophers, not scientists.[2] The scientific approach
spurned a millennium-old tradition and was fraught with danger. When scientists sought answers in the heavens rather than

2. It took a while for the term *natural philosopher* to be replaced by *scientist*.

in the writings of the ancients, they wandered into dangerous territory. In the church's eyes, the greatest affront was when the first scientific cosmologists stated that observation and calculation, rather than divine revelation, could reveal the workings of the heavens. Scientists posed a direct threat to the shepherds of the Christian flock. Ironically, the first shot of this first cosmological revolution began with Copernicus, a devout clergyman.

Copernicus was not an astronomer by trade; he was a doctor. Doctors of the day had to be skilled in astrology, the better to deduce ailments and rebalance the body's humors. (Medieval medicine was another descendant of ancient Greek wisdom.) However, when Copernicus used Ptolemy's clock-works to prepare his astrological charts, he found that Ptolemy's cosmos seemed complicated, cumbersome, and unsatisfying. The skilled doctor spent much of his life trying to come up with a cleaner, simpler explanation of the motions of the planets.

It was a difficult task. The Ptolemaic system did an ad-mirable job of explaining the complicated, backward-and-forward motion of the five known planets: Mercury, Venus, Mars, Jupiter, and Saturn. It seemed like the clockwork of the heavens was inherently complex. Any system that obeyed the Aristotelian laws of physics, which demanded that Earth be at the center of the universe, would necessarily have a Byzantine mess of circles within circles to make it work.

Copernicus came up with a radical solution. He eventu-ally realized that the Ptolemaic system was exceedingly com-plicated because it stuffed Earth into the center of the universe where it does not belong. By putting the sun at the center instead, and by allowing the planets to move around the sun rather than the Earth, Copernicus reduced the num-ber of epicycles, the wild gyrations of the planets, from about eighty to about thirty. Copernicus's sun-centered system was cleaner and simpler, but it was not perfect. Actually, the Ptolemaic, Earth-centered system was more accurate at pre-dicting the motions of the planets. If scientists had had to

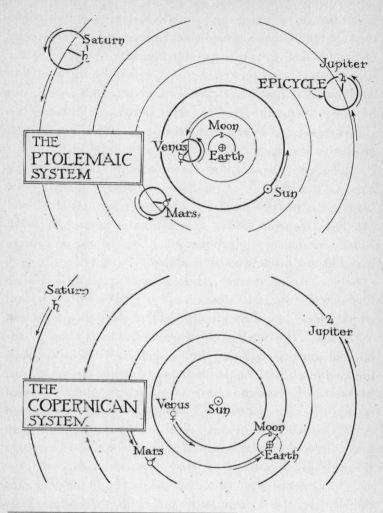

The Ptolemaic and Copernican cosmologies

choose a single system based solely upon the quality of its predictions, they would have chosen Ptolemy's, even though the ancient Greek's clockworks were so much more complex than the Polish doctor's.

Nonetheless, the Copernican system was the first rumble of an approaching storm. By putting the sun rather than the

Earth at the center of the universe, Copernicus challenged the very foundation of Aristotelian cosmology, just as Nicholas of Cusa had. In Copernicus's universe, the Earth was in the heavens, just as all the other planets were; the whole idea of earth and water sinking to the center of the cosmos could not be true if the Earth were floating up in the sky. Perhaps even Cusa's wildest idea was true, that each star in the sky was home to alien worlds. But in Copernicus's day, the case for the heliocentric system was not yet airtight, and the church had not caught scent of the danger it posed. Indeed, when Copernicus in 1543 published his great work, *On the Revolutions of the Heavenly Spheres*, he dedicated it to Pope Paul III. But Copernicus was a prudent man. He took the precaution of publishing it while on his deathbed.

By the time Copernicus died, another upheaval—this one theological—was well under way. In 1517, Martin Luther nailed ninety-five theses to the door of the castle church at Wittenberg. His hammer blows resounded throughout Christendom as more and more people, angry with the corruption of the church, renounced their allegiance to the pope. The Protestant Reformation was born, and it quickly gained strength. To counter the threat, the church trained an elite corps of intellectual clerics, the Jesuits, who would be ideal ground troops for the war on Protestantism. Jesuit theology was strongly dependent on Aristotelian ideas; they used the ancient Greek wisdom to explain the motion of the planets in the sky and the transmutation of bread into Christ's body during the sacrament of Communion. Aristotle became a potent weapon in the church's intellectual arsenal. To attack Aristotelianism became tantamount to challenging the word of the Bible and the holiness of the Communion.

The church, under attack, was less and less able to allow challenges to Aristotle. Yet it was more than half a century before the church banned Copernicus. (Luther saw the problem before the Catholic Church did and rushed to denounce Copernicus as a fame-seeker. "People give ear to an upstart astrologer who strove to show that the earth revolves, not the heavens or the firmament, the sun and the moon," he wrote. "This fool wishes to reverse the entire science of astrology; but sacred Scripture tells us that Joshua commanded the sun to stand still, not the earth." That passage in Joshua caused no end of trouble.)

As the Protestant Reformation grew, the church became harsher and harsher with its critics, including those who took up the cause of a sun-centered cosmology. Giordano Bruno learned this the hard way. On February 17, 1600, after Bruno had endured a long imprisonment, the church burned him at the stake for his heretical ideas. Bruno had embraced Copernicus's heliocentric model of the solar system, and like Nicholas of Cusa he declared that the Earth was only one of an infinite number of worlds.

Nobody knows what role Bruno's cosmology played in his condemnation. The records of Bruno's trial at the hands of the Roman Inquisition are lost to history, so it is unclear whether he was burned for his cosmology, his personal conduct, or both. But the crackdown on heresy was getting ever stronger, even as a noseless noble (with a pet midget) and a German astrologer-mathematician turned the Copernican system into an instrument with clockwork precision.

The noble, Tycho Brahe, was a sybaritic Dane. Born in 1546, he was a glutton. (Overeating led to his death half a century later.) For his amusement, Brahe kept a dwarf whom he fed with table scraps, but this was not nearly as unusual as Brahe's physical appearance. Brahe lost much of his nose in a duel—he was a better astronomer than fencer—and had a silver prosthesis. Yet this comical character would strike blow after blow to Aristotle's perfect universe.

On a chilly November night in 1572, Brahe spotted a new star in the constellation Cassiopeia. We now know that he had seen a supernova, the spectacular death throes of a doomed star, but to Brahe it was an incredible paradox. The Aristotelian heavens were supposed to be perfect and immutable, yet they had changed before his eyes. Within a year or so, Brahe had enough data to show that the new star was quite distant, farther out than even the moon, so it was clearly not an atmospheric phenomenon. The new star—the imperfection—was part of the supposedly unchanging heavens.

Then Brahe set up the best astronomical observatory of his day, Uraniborg, located off the coast of Copenhagen. It was an enormous undertaking; the sextants, quadrants, and other instruments (telescopes had not yet been invented) cost the Danish government about a third of its national income. It was worth every penny. In 1577, Brahe showed that comets, the irregular fuzzy bodies that appeared in the sky from time to time, were also more distant than the moon, so they were heavenly bodies rather than luminous clouds in the atmosphere. He also detected a slight periodic variation in the speed of the moon's orbit around the Earth. It was blazingly clear: the heavens were fickle and imperfect.

However, Brahe's lasting legacy came from his reams of observations. In 1600, he convinced a young astrologer and mathematician named Johannes Kepler to join him in Prague as an assistant. (Brahe had moved from Uraniborg after arguing with the king of Denmark.) Kepler used Brahe's data to show that the planets didn't move in perfect circles.

Unlike Brahe, Kepler believed in Copernicus's sun-centered theory, having learned it from his mathematical mentor in school. Kepler seemed to be attracted to the simplicity of the heliocentric universe, even though it was still less accurate than the ancient Ptolemaic, geocentric cosmology. Kepler fixed this defect in 1609 when he announced that the planets move in ellipses rather than circles. After years of tedious labor, Kepler broke out of the circular universe imposed by Ptolemy and by Copernicus. Everything fell into place, and the heliocentric universe, freed from all the philosophic preconceptions that held it back, described the motions of the planets more accurately than did the Ptolemaic system. Heliocentrism was simpler, more accurate, and more elegant than geocentrism. It was the death knell for Ptolemy and Aristotle, and for the cosmology that underpinned the theology of the church.

The church finally was fully awake to the danger posed by the new philosophy, and everyone who threatened Aristotle's framework was, himself, in mortal peril. Even Galileo Galilei, friend of Pope Urban VIII, was in danger of being burned at the stake. In 1609, the same year that Kepler published his *New Astronomy*, Galileo heard that a Dutch lensmaker, Hans Lippershey, had created a device to make distant objects look nearer. Galileo immediately built himself one of these new instruments—the telescope—and turned it to the heavens. Everywhere he looked, like Brahe before him, he saw evidence that Aristotelian cosmology was wrong. His discoveries were systematically destroying what was left of the Aristotelian universe.

When Galileo looked at the moon, he saw mountains and craters. A heavenly body, which, according to Aristotle, was made of purer stuff than the Earth, was just as pitted and scarred as the craggiest parts of our own planet. When he looked at the sun, he saw blotches—sunspots—that belied the orb's perfection. Turning the telescope to Jupiter, Galileo found four bodies that orbited the giant planet; here was incontrovertible proof that not everything orbits the Earth. If these distant moons circled Jupiter, ignoring Earth, it was hard to imagine that the Earth was truly the center of the cosmos. Looking at Venus, Galileo noticed that the planet went through phases, waxing and waning like the moon. This was predicted by the Copernican system (indeed, Copernicus realized that the apparent lack of phases was a problem for his theory) but nearly impossible to explain within Aristotelian-Ptolemaic cosmology.

The telescope was the big gun of the first cosmological revolution, and Galileo wielded it with skill, shooting down one Aristotelian conceit after another. Galileo's observations convinced him that Aristotle was wrong, and Copernicus was right. Unfortunately, this made his scientific investigations a theological matter.

In 1613, Galileo wrote to a student of his, a priest, arguing that if the Bible seemed to contradict scientists' observations of the workings of nature, then the interpretations of the Bible must be mistaken. To Galileo, science was stronger than theology; if the two contradicted each other, then theology must give way, not science. This was heresy. From the church's point of view, Galileo was trying to reshape Christian theology, replacing the Aristotelian philosophy at its center with a new and uncertified doctrine. Galileo was no Saint Thomas Aquinas — he had no right to dictate theology to the church. Galileo was on the edge of becoming a heretic.

In 1616, Cardinal Bellarmine, head of the Roman Inquisition, called Galileo into his office. The cardinal warned Galileo that Copernicanism was heresy and, according to one account, told Galileo not "to defend or hold" Copernican theory. Galileo took the warning seriously because the church was becoming more and more brutal in persecuting heretics. On December 21, 1624, three years after the death of Bellarmine, a crowd gathered in Rome to watch the immolation of the body of a heretic who had died three months earlier. Even the dead were not safe from the flames of the righteous.

Unfortunately for Galileo and fortunately for posterity, the scientist in Galileo could not stay away from the new cosmology despite the increasing danger. He was irresistibly drawn to it by his own observations, and he continued to write about the new science of the heavens. In 1633, the Inquisition condemned Galileo as a heretic.

"We say, sentence and declare that you, Galileo . . . ," the condemnation reads, "[have] believed and held the doctrine, false and contrary to sacred and divine Scripture, that the Sun is the center of the world and does not move from east to west and that the earth moves and is not the center of the world; and that an opinion may be held and defended as probable after it has been declared and defined to be contrary to Holy Scripture." In other words, Galileo had claimed that science could force theologians to change their views, rather than vice versa. Heretics who confessed and recanted could escape with their lives; those who stuck to their incorrect ideas were burned. Galileo prudently recanted and was imprisoned rather than burned. As a favor to his old friend, Pope Urban VIII allowed Galileo to spend his perpetual imprisonment at his own home, rather than in a dank cell in the Vatican.

The church was provably, criminally wrong. The church had condemned an innocent man. It punished Galileo for the revolution that destroyed its wrongheaded cosmology. But it held its ground for centuries. In 1930, Pope Pius XI canonized Bellarmine.

Even today, the Catholic Church is struggling with its past—not very successfully. In 1992, Pope John Paul II expressed regret that the Galileo incident came to represent the "myth of . . . the Church's supposed rejection of scientific progress," even though it was merely a case of "tragic mutual incomprehension" between Galileo and the church. Mistakes were made. But the church was not alone in its error, according to the Vatican. Cardinal Paul Poupard defended the inquisitors, stating that Galileo's arguments were not airtight. "In fact, Galileo had not succeeded in proving in an irrefutable fashion the . . . motion of the Earth," he said in 1992. "It took more than 150 additional years to find the optical and mechanical proofs of the Earth's motion." Galileo too was at fault, for not making the case sufficiently, according to the Vatican.

Nonetheless, it is clear that the church had erred. It clung to a doomed cosmology. Galileo was right; Aristotle was wrong. When Isaac Newton formulated the laws of motion and of gravitation, mathematicians and physicists could derive Kepler's laws, and even the motion of the solar system, from two simple equations. Plug in the masses of the planets and the sun, give their initial positions and velocities, and they could calculate, to great precision, where any given planet would be in the sky at a future date.

Despite the Inquisition's judicial triumph over Galileo, the first cosmological revolution had unseated nearly two millennia of philosophy and theology and replaced it with science. Much to the chagrin of the church, when science contradicted theology, theology had to change. In 1822, the Catholic Church finally removed Copernicus's *On the Revolutions of the Heavenly Spheres*, Kepler's *New Astronomy*, and Galileo's *Dialogue Concerning the Two Chief Systems of the World* from the Index of Forbidden Books. The church accepted the new cosmology that had smashed the nutshell universe; indeed, the clergymen began to explore it, eventually founding their own observatory.

Run by Jesuits, the Vatican Observatory now embraces what the church—and the Jesuits—once rejected: the scientific method. When the current director of the Vatican Observatory, Father George Coyne, SJ, saw me examining the portraits of Galileo and Bellarmine on the wall of the Vatican, he pointed to a third portrait on the same wall, also wreathed with laurels and crowned by the keys of heaven. It was a picture of Cardinal Baronius, whose famous saying Galileo vainly used in his defense: "The Bible teaches us how to go to heaven, not how the heavens go."

Chapter **3**

The Second Cosmological Revolution

[HUBBLE AND THE BIG BANG]

The universe, as has been observed before, is an unsettlingly big place, a fact, which, for the sake of a quiet life, most people tend to ignore. Many would happily move to somewhere rather smaller of their own devising, and this is what most beings, in fact, do.

— DOUGLAS ADAMS, *THE HITCHHIKER'S GUIDE TO THE GALAXY*

The new cosmos described by Copernicus, Kepler, and Galileo was much more vast than Aristotle's, and much more frightening. No longer was the Earth at the center of the universe; it was one of a multitude of worlds, each perhaps teeming with bug-eyed alien beasties. Nevertheless, by modern standards Galileo's universe was very small indeed.

Three centuries after Galileo, a second cosmological revolution forced scientists to accept just how mindbogglingly large the universe really is. Confronted by a new set of observations in the 1920s, cosmologists were forced to admit that their old model of the "universe" encompassed just one little patch of stars among millions and millions of galaxies throughout the cosmos. It was incredibly unsettling to realize just how insignificant our planet is compared with the vastness of space.[1]

As minuscule as the Earth, the sun, and the Milky Way are with respect to that vastness, the most shocking part of the second cosmological revolution wasn't the size of the universe. At the same time that cosmologists figured out just how large the cosmos is, they realized that it is unstable. They saw that the universe is not eternal and unchanging; it is finite. It was born and it will die.

The second cosmological revolution forced scientists to confront the creation of the universe and its demise. This was such an unpleasant idea that some physicists scrambled to find a way out of the demise of the universe. Even straight-laced Albert Einstein worried about "being confined in a madhouse" as he made a desperate attempt to stave off the impending death of the cosmos. Einstein took the risk because he knew that the second cosmological revolution would force astronomers to look directly at the face of creation.

The first cosmological revolution, the revolution that shattered Aristotle's neat little nutshell, gave scientists a powerful new theory of the universe. That new theory went hand in hand with a new astronomical tool: the telescope. Galileo would never have seen the imperfection of the heavens without the aid of his little tube with lenses. In the years after Galileo's death, astronomers kept improving the power and quality of their telescopes, but it was another three hundred

---

1. In Douglas Adams's *Hitchhiker's Guide to the Galaxy* trilogy, the ultimate torture machine is known as the Total Perspective Vortex—a diabolical creation that can turn any living being into a raving lunatic. Inside the chamber is a picture of the entire vastness of the cosmos with a tiny arrow that reads, "You are here."

years before telescopes were powerful enough to ignite another intellectual wildfire.

Of course, the intervening three hundred years were not a waste by any means. Physics and astronomy flourished, changing science's view of the workings of the universe. Newton and Aristotle assumed that light traveled instantaneously—it spent no time in moving from the sun or the stars to Earth. In 1676, Danish astronomer Ole Rømer showed that Newton was wrong. Light traveled at a finite speed. The discovery came when Rømer used Newton's laws to calculate the orbit of Io, one of Jupiter's moons. As he charted the motion of the fuzzy little dot in the sky, he realized that the orbital positions he observed were a tiny bit off. What's more, how far off the observations were depended upon how far Earth was from Jupiter. Rømer realized that the discrepancy between theory and observation was not the fault of an incorrect theory; it occurred because light took a few minutes to bridge the enormous distance between the heavenly bodies.[2]

The three hundred years after Galileo were an age of exploration. Just as surveyors on Earth were busily mapping out the size of the continents, astronomers were attempting to map the distances between the planets and the stars. But unlike earthbound geographers, astronomers can't travel to Jupiter and set up a surveyor's transit to measure precise distances. Astronomers had to rely on a different set of tools to map the heavens. For many years, the most powerful tool in the celestial surveyor's toolbox was *parallax.*

Parallax might seem a little bit tricky to understand at first, but actually it's right in front of your nose. Literally. To see the effect of parallax, simply face a distant object and hold your index finger near the tip of your nose. Close your left eye and keep the right one open. Now switch, closing the right

---

2. The speed of light, *c,* is 186,282 miles per second. While this seems extremely fast, space is so vast that it takes light a bit more than four years to travel from the nearest star to Earth. Such huge distances are measured in *light-years,* where one light-year is the distance that light can travel in one year: approximately six trillion miles.

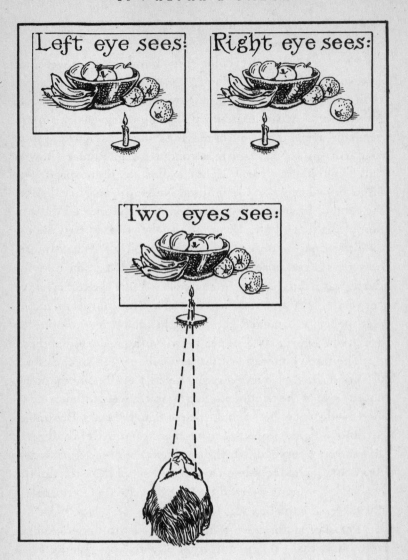

Parallax

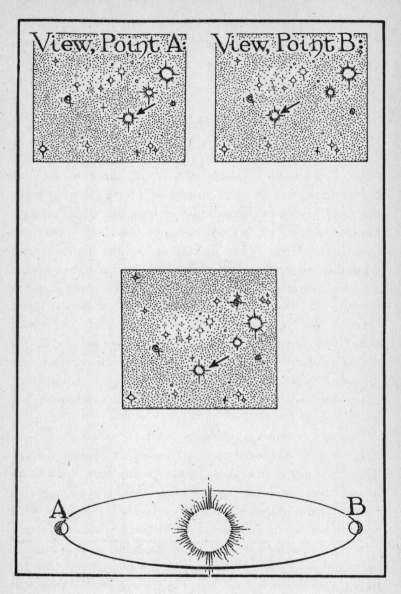

Parallax in astronomy

one and keeping the left one open. Switch. Switch again. Switch again. You should see the finger appear to move back and forth about an inch or so. Now move the finger farther from your nose, perhaps at arm's length. When you alternately close each eye, you will see that the finger seems to move a bit less. The farther away your finger is from your eyes, the less your finger will move from side to side. This is parallax.

Parallax works because your eyes are a small distance apart. This means that each has a slightly different image of the world. Both eyes agree, in general, about the relative positions of objects, such as things in the distant background, but there are subtle differences between each eye's perspective. The closer an object is to your face, the more extreme those differences are. For example, when your right eye sees a finger close to the tip of your nose, it perceives the finger at the extreme left of its field of view. From your left eye's perspective, the finger is at the extreme right side of the world. When you alternately close your left or your right eye, you can switch between one perspective and the other. The finger jumps with respect to the background, because the eyes disagree about whether the finger is at the extreme left or the extreme right.

It's usually not a good idea to have one part of your brain disagree with another, but in this case your brain uses the disagreement to its advantage. By comparing how much those images disagree, it can estimate how distant your finger is. The more the images clash with each other, the closer your finger must be. As you move your finger farther and farther away, the two images agree more and more, and eventually your brain is unable to distinguish the two images. But for objects that are relatively close, two images, viewed from two perspectives at some distance from each other, can reveal how far away an object is.

Parallax also works in astronomy. When astronomers make two observations at some distance apart (since the Earth moves around the sun, waiting six months automatically provides a different vantage point on the other side of the solar system), they can figure out how distant an astronomical object is. In the early nineteenth century, mathematician and astronomer Friedrich Bessel used parallax for the first time, measuring the distance to the star 61 Cygni (about ten light-years).

Unfortunately, Bessel and his successors could not use parallax on most of the objects in the universe. Earth's orbit is so small compared with the distances between the stars that astronomers could only use this method on the very closest objects. Nonetheless, they began to map our celestial neighborhood, figuring out the distance to the planets, the sun, and a few hundred nearby stars. They also discovered that the ribbon of light in the sky, the Milky Way, was a vast disk-shaped collection of stars, and they even had a fairly good idea where Earth was within that disk. However, they could not figure out the distances to the mysterious, luminous "nebulae" that dotted the heavens.

In the years before the French Revolution, astronomer and comet hunter Charles Messier made a catalog of fuzzy objects in the heavens that could easily be mistaken for comets. To Messier, his catalog of glowing clouds, luminous spirals, and other nebulous objects was merely a device to help him find comets more easily. But to other astronomers, Messier's objects, along with similar nebulae found by other astronomers, became a fuzzy enigma. What were these mysterious clouds? Were they nearby clumps of luminous gas, or were they distant collections of stars?

Astronomers argued and spilled gallons of ink in scientific journals, but could not come to any conclusion. They had no way to gauge the distance to those nebulae—the parallax method just couldn't measure distances that large—so their arguments were fruitless. The bickering reached a climax on the evening of April 26, 1920, when astronomers Harlow Shapley and Heber Curtis squared off in the National Academy of Sciences building in Washington, D.C. Shapley argued that nebulae were nearby gas clouds. Curtis insisted that they were distant galaxies of stars, like the Milky Way. But the "Great Debate," as it came to be known, was much more than an argument over gas clouds. It was a fight over the size of the universe.

Shapley's universe was little different from Kepler's. Earth, a lowly lump of matter, circled its sun. The sun was part of the Milky Way, which encompassed the whole known universe. Everything in Shapley's cosmos was contained in a flattened disk of stars, ourselves included. While this view of the universe was much, much broader than Aristotle's nutshell cosmology, it was still a relatively small place, a few tens of thousands of light-years across. Curtis's cosmos, on the other hand, was much, much larger than Shapley's. The Milky Way, instead of comprising the entire known universe, was merely one luminous pinwheel among thousands and thousands of similar whorls of stars. The universe was filled with countless Milky Ways, each as grand as our own galaxy. Curtis's cosmos was millions, or even billions, of light-years wide, at the very least. Maybe it was infinite.

In a sense, though, the Shapley-Curtis debate was almost as useless as arguing about angels and pins. In 1920, neither man had the tools to settle the question; neither knew any way to gauge the distance to those mysterious nebulae. When the debate ended and the audience filtered slowly out of the auditorium, nobody had any idea whether Shapley or Curtis had won the argument. Either scientist could be correct: the universe could be vast, or it could be a relatively small clod of stars. By the end of the decade, a young astronomer named Edwin Hubble would settle the question once and for all. Using an enormous telescope at Mount Wilson, near Pasadena, California, Hubble struck two huge blows for cosmology. Hubble's discoveries forced scientists to grapple with the immensity of the universe—and forced them to look at the very moment of creation. Hubble would start a second cosmological revolution and blow away the idea of a cozy little universe.

＊　＊　＊

Hubble needed a weapon to start a revolution. The heart of that weapon was delivered on July 1, 1917, when a gigantic plate-glass mirror, one hundred inches across, arrived atop Mount Wilson. The four-and-a-half-ton mirror was to become the core of the world's largest telescope.

A telescope, like your eye, is a device to collect light. When you look up at night and see a distant star twinkling in the sky, your eye is telling your brain that it has gathered a handful of photons from that star. The lens of your eye bends those photons so they strike your retina—the surface at the back of your eye that detects light—and the retina, in turn, sends a signal that your brain interprets as a point of light. Like all instruments, the eye is imperfect. You don't see all, or even most, of the stars in the heavens. Your eye isn't sensitive enough to detect them. The retina doesn't detect every photon that strikes it; it misses a pretty large proportion of the light that enters the eye, and as a result, the dimmer stars, the ones that don't have many photons streaming down from the heavens, are invisible to the unaided observer. With a telescope, however, you can scoop a lot more light into your eye and see many, many more stars.

When Galileo pointed his telescope at the heavens, his tube full of lenses gathered a great deal of light from a small portion of the sky and concentrated it on his retina. When he looked at Jupiter, his telescope captured enough light for him to see four tiny specks of light around the giant planet. When he pointed the telescope at the sun, he focused such an enormous amount of light on his retinas that he damaged them—and eventually went blind. As engineers built ever bigger lenses and mirrors, astronomers' telescopes became better and better light gatherers, and astronomers saw finer and finer details of objects in the heavens. For example, in 1763, Charles Messier spotted a fuzzy blob to which he assigned the number thirteen. His journal describes it as "a nebula without a star, discovered in the belt of Hercules; it is round and brilliant, the center is more brilliant than the edges." With his telescope, which was about eight inches in diameter, Messier saw M13 as a glowing cloud without any individual specks, which is why he described it as a starless nebula. But by 1833, British astronomer John Herschel described it as a "very rich cluster" of stars, "of which there must be thousands." With his more powerful telescope, Herschel saw that Messier's "starless" fuzzy blob was really a collection of thousands and thousands of stars, too small to be resolved with Messier's little telescope. The bigger telescopes got, the more astronomers learned about the objects in the heavens.[3]

When the First World War ended, the hundred-inch mirror on Mount Wilson was the best light-gathering device in the world. In 1919, a young Edwin Hubble joined the Mount Wilson observatory, and he was able to crack the mystery of how far away the fuzzy nebulae are. It was not an easy question to answer, since their great distances rendered parallax useless. Luckily, there was another method.

Imagine that you are trying to judge the distance to an obelisk. All obelisks look pretty much the same, from the relatively small, like Cleopatra's Needle in London, to the enormous, like the Washington Monument. If someone hands you a photograph of an obelisk in the distance, you might not have any sense of scale—you wouldn't know whether the obelisk was big or small, or whether the photographer was standing close to the obelisk or far away. However, if a person is standing in the photograph next to the obelisk, you can figure it out. Since people are all roughly the same size, a person in the photo can help you estimate the size of the obelisk and how far away the photographer stood while taking the picture. If the person is very, very tiny compared to the obelisk, then the monument has to be quite large and distant; if the person is

3. Astronomers got another boost by replacing the astronomer's eye at the end of the telescope with a device much more sensitive than the retina: the photographic plate. Modern telescopes now use an electronic eye called a charge-coupled device, which is better still.

large compared to the obelisk, then the needle has to be relatively small and close by. Your brain automatically uses the person, an object of known size, as a measuring stick to gauge the height of the obelisk and the distance from the photographer. In astronomical terms, the person is a *standard ruler*.

If Hubble had found a standard ruler in one of the mysterious nebulae, he would have been able to figure out how distant the nebula was. Unfortunately, Hubble didn't. (Standard rulers have recently become important in cosmology, but astronomers in the 1920s didn't know the sizes of any objects visible at astronomical distances.) However, Hubble had something just as good: a standard candle. Just as a standard ruler is an object of known size, a standard candle is an object of known brightness that can be used to judge distance in a similar way. (Our brains measure brightness less well than they measure size, so we're much less likely to use standard candles in real life. However, by using a photographic plate, or an electronic equivalent, astronomers have made it a very sensitive technique.)

If you give a person a flashlight and have him walk away from you, the flashlight will seem to get dimmer and dimmer as the person trudges off into the distance. Conversely, if you know how bright the flashlight's beam is, you can estimate how far away he has gone—the dimmer, the farther away. So if a person standing next to an obelisk shines a flashlight of known brightness, you have a second method of judging the distance to the obelisk. That is a *standard candle*.

In Hubble's day, astronomers had a standard candle in a class of stars called Cepheid variables. Cepheid variables have the peculiar property of dimming, then brightening, then dimming again, puffing up and collapsing and puffing up in a seemingly endless cycle. These moody Cepheids are useful because their brightness is related to the speed of their cycles. By measuring how quickly a Cepheid goes from bright to dim and to bright again, astronomers can figure out just how bright the Cepheid is at its peak. As an object of known brightness, a

Standard rulers and standard candles

Cepheid variable star becomes a standard candle. Find one, and you can figure out how far away it is.

Astronomers measured the distance to a handful of nearby Cepheids, but the real prize came in the early morning of October 6, 1923. The hundred-inch telescope at Mount Wilson was pointed to the "Andromeda nebula," the king of the fuzzies in the sky. It took a photographic image of the nebula, and the enormous mirror of the telescope captured something that nobody had seen before. A star had brightened and dimmed . . . and then brightened again. At first Hubble, who saw it brighten and dim, thought it was a nova, a bright explosion that flares and dims away. But novas don't brighten again. Hubble crossed out the "N" on the photographic plate and wrote "VAR!" It wasn't a nova; it was a Cepheid variable star in the Andromeda nebula.

Hubble had found a standard candle and could use it to gauge the distance to that glowing cloud. Hubble's calculations showed that the Cepheid was hundreds of thousands of light-years away, so distant that the light reaching Hubble's photographic plate was hundreds of thousands of years old. This was far beyond the outer reaches of our galaxy. The Andromeda nebula was, clearly, not a clump of matter inside the Milky Way. It was a galaxy unto itself, a collection of stars perhaps as vast as our own galaxy. Hubble found more Cepheids in Andromeda, confirming his conclusion: Andromeda was a separate galaxy, hundreds of thousands of light-years away.[4] (The object had to be renamed the Andromeda galaxy instead of the Andromeda nebula.)

Then Hubble turned the telescope to other swirling disks, and much to his delight, those disks also had Cepheid stars. It turns out that Andromeda, as far away as it is, is just the closest of a countless host of galaxies. Each of those spiral nebulae is an entire galaxy like our own, vastly more distant than

---

4. The modern value is a bit more than two million light-years away. Hubble made a slight error that led him to underestimate the distance to Andromeda. The error is described more fully in chapter 4.

cosmologists had imagined. The Great Debate was over, and Curtis was right. The Milky Way is just one island in the enormous ocean of space. Shapley's tiny, one-galaxy cosmos was simply wrong. Instead of being a few tens of thousands of light-years across, the universe had to be millions (and more) light-years wide. Not only are there other suns, there are other galaxies, vast collections of suns.

Hubble's cosmos was a humbling, empty place, broken only by small oases of stars. This discovery was monumental, but surprisingly it was the lesser of his two discoveries. His distance measurements made cosmologists realize just how vast the universe is. But in 1929, he forced them to ponder the birth of the universe—and its death.

The second of Hubble's discoveries relies upon the "fingerprints" of the stars. All stars (and not just the Hollywood variety) are essentially balls of hot gas. Hot gas emits light of various colors; in fact, each has a distinctive color. When you look at a sodium light, it looks yellowish, while a neon sign looks red. If you use a prism to split the light into its components, you will see a series of stripes of different colors that reveals what gas you are looking at; each gas, be it sodium, neon, hydrogen, or helium, has a different, unique, set of stripes. These stripes are as distinctive as fingerprints. Astronomers can look at the stripes in a star's light spectrum and figure out precisely what elements it is made of (and even the relative abundances of those elements).

When Hubble used this trick, passing the light from other galaxies through a prism, he saw their chemical fingerprints—particularly hydrogen, as hydrogen is, by far, the most common element in the universe. Oddly, he noticed that the stripes were not in quite the right places; the colors were slightly off. The relative positions of the stripes in the fingerprint were correct, but they were all shifted a bit toward the red end of the spectrum. Because astronomers had already seen that sort of shift before, Hubble quickly figured out what was happening. He was seeing what is known as a redshift, an example of the Doppler effect, the same phenomenon that state troopers use to nab speeders.

When a train rushes toward you and blows its whistle, you can hear the Doppler effect in action. As the train passes, all of a sudden its whistle drops from a high pitch to a low pitch. This occurs because the train's motion crushes and stretches the sound waves of the whistle. Ahead of the engine, the sound waves get squished together, making the whistle higher in pitch than normal; behind the engine, the train's motion stretches out the sound waves, making the whistle low-pitched in its wake. This squashing and stretching is the Doppler effect. When a police officer zaps a car with a radar gun or a laser beam, he is really measuring how much the motion of the car is compressing the reflected radiation. By measuring that squashing, he can figure out how fast the car is moving, and give the driver a $250 ticket. Isn't science wonderful?

Radar and laser guns shoot beams of light, so it should come as no surprise that the Doppler effect pertains to light as well as sound. One important difference between light and sound, however, is what gets shifted. With sound, a shift in frequency means a change in tone: the higher the frequency, the higher the pitch. With light, on the other hand, a shift in frequency means a change in color. The higher the frequency, the bluer the light. While a train moving toward you will have a higher-pitched horn than normal, a star zooming at you will have bluer-than-usual light. Conversely, a train speeding away has a lower-pitched sound, and a star receding from you will look redder than usual.

When Hubble found that the stellar fingerprints were redder than usual—they were redshifted—he realized that the galaxies he was seeing were moving away from the Earth.[5]

---

5. Big bang doubters take issue with the idea that the redshift of the galaxies' light means that they are moving away. Instead, some say that light gets "tired" and red as it streams across long distances, so the reddening is due to "tired light" rather than the Doppler effect. This idea has been thoroughly debunked. See appendix A for a fuller accounting.

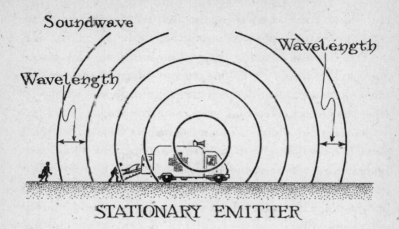

Soundwave

Wavelength

Wavelength

STATIONARY EMITTER

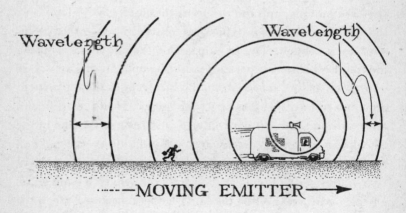

Wavelength

Wavelength

·----MOVING EMITTER----→

The Doppler effect

Hubble was surprised to realize that *all* the galaxies were zooming away. Worse yet, the farther away the galaxy was, the faster it was speeding away. This meant the universe is flying apart! It was a shocking conclusion, and it was the first salvo of the second cosmological revolution, and it would destroy the idea of an eternal, unchanging universe.

How could anyone explain the bizarre behavior of these galaxies? We can turn to modern cosmology for the answer. Imagine that the universe is a rubber balloon, and the galaxies are like tiny polka dots on its surface. As the balloon blows up, the polka dots all rush away from one another. From the perspective of any particular dot on the surface of the balloon, all the other dots are speeding away, and in fact, the more distant ones are rushing away faster than the nearby ones. This is roughly equivalent to what is happening in the cosmos. The universe is expanding, and we in the Milky Way see that expansion because galaxies in all directions are zooming away from us. The farther away they are, the faster they are moving, and the redder their spectra.

This picture of an expanding universe presents a philosophical problem. The expansion of the universe is like a movie of a balloon expanding at a certain rate (a rate that is expressed as the *Hubble constant,* which relates the distance of a galaxy to how fast it is speeding away). However, you can also run a movie backward. If we could reverse the film of the universe's expansion backward, you would see the balloon shrink and shrink and shrink at a given rate over one billion, two billion, ten billion years . . . and then what? It can't go on shrinking forever. When the film reaches a certain time in the past (now thought to be about fourteen billion years ago), the balloon must shrivel completely, shrinking into a point and disappearing. The balloon has collapsed entirely. Fourteen billion years ago, the balloon universe was completely shrunk. Earlier than that, the universe could not have existed in its present form.

To someone watching the movie in its proper direction,

the universe would seem to spring to existence from that tiny point. This is the big bang. The universe had to have a beginning, and it wasn't infinitely long ago. The universe is only fourteen billion years old; before that, the universe simply did not exist. This idea—that the universe must have a birth date—flabbergasted and repulsed many scientists of the day, including Albert Einstein. When Einstein formulated the general theory of relativity, which in effect describes the nature of the "rubber surface" of our balloon universe, he quickly realized that the universe his equations describe was necessarily unstable. There was no way, if Einstein's theory was correct, that the universe could go on, changeless, eon after eon. The theory of relativity said that the universe had to be either expanding or contracting; it could not stay the same.

Einstein found the idea of a changing universe so abhorrent that he tinkered with his equations. To "fix" the problem of a finite universe, he added a term, a constant denoted with a Greek capital letter lambda (Λ), which balanced out the forces acting upon the rubber-sheet fabric of the cosmos, making it stable once more. Einstein's *cosmological constant* was a way of avoiding the consequences of an ever changing universe, a universe with a beginning and an ending. There was no scientific justification for adding Λ to the equations, and it seemed rather nonsensical (Einstein jested about being committed to an insane asylum for proposing it), and he would soon regret his tinkering. When Einstein learned of Hubble's discovery, something that Einstein's own equations could have predicted, he called the cosmological constant the greatest blunder of his career.

Other scientists also rejected the idea of a finite universe with a birth, and perhaps a death. In 1948, astronomers Hermann Bondi, Thomas Gold, and Fred Hoyle proposed an alternative to the big bang theory, which gave a glimmer of hope to those who wanted an eternal universe. The so-called steady-state theory was based upon the idea that the universe as a whole stays the same, even as the individual galaxies move away from one another and die. In such a steady-state universe, matter and energy eternally rushes forth from fountains, condensing into galaxies. The newly formed galaxies then speed away from one another. This eternal, constant creation of matter avoids the need for a cataclysmic birth.

Hubble's discoveries ushered in a new question about the nature of the universe: big bang or steady state? Is the universe finite, or is it eternal and unchanging? The debate raged for decades. Cosmologists leaned toward one or the other depending on whether they found the idea of a finite universe comforting or disturbing. But the next skirmish in the second cosmological revolution would answer the question once and for all. Hubble's discovery forced cosmologists to ponder the birth and death of the universe, but a ubiquitous glow in the sky known as the cosmic microwave background gave them their first glimpse of the fiery birth of the cosmos. The ancient comfort of an eternal, unchanging universe was gone forever.

The universe has walls of fire. No matter where astronomers point their telescopes, they see a distant sheet of light surrounding us. Beyond that enormous wall of radiation, more distant than the most ancient stars and galaxies, astronomers can see nothing. We are caged in by this surface: the cosmic microwave background (CMB), the faint afterglow of the big bang.

The big bang was a massive explosion that created all the mass and energy in the universe, as well as the fabric of spacetime. This fabric, which is described by Einstein's general theory of relativity, inflated rapidly after the cataclysm, but within a tiny fraction of a second, the rapid inflation slowed down and freely roaming subatomic particles, quarks, began to form protons and neutrons, which were constantly being buffeted about by incredibly intense and energetic light. As objects expand, they cool (feel the nozzle of a propane tank after you have let its gas expand into a backyard grill) so the ever growing universe cooled down. The intense light that suffused the universe stretched out and became less energetic; after a few minutes, the temperature of the newborn cosmos dropped. Some of the protons and neutrons coalesced into nuclei of deuterium (hydrogen with an extra neutron), helium, and a few other heavier elements. The universe was filled with atomic nuclei and electrons — and with light. Whenever an electron tried to combine with a nucleus, it was struck by a photon (a particle of light) soon afterward, stripping it away; conversely, a photon could not get very far before it scattered off an atom trying to coalesce. Light was trapped in a cage. This was the nature of the universe until about 400,000 years after the big bang, when the expanding universe cooled enough for another change: electrons combined with their nuclei once and for all. This *recombination* freed light from its confines. The entire universe glowed brilliantly.

The universe continued growing and never stopped. The light from recombination still rattles around the cosmos, but as the fabric of spacetime expanded, so too did the light. Over billions of years, the ultra-high-energy gamma rays stretched into x-rays, visible light, and now, fourteen billion years after recombination, microwaves. The scream of light has become a mere whisper, a faint glow with a temperature of 2.7 degrees above absolute zero. This is the cosmic microwave background, also known as the cosmic background radiation. This radiation comes from every direction in the sky.

Though modern cosmologists now know of the existence of the cosmic microwave background, in the decades after Hubble's observation nobody had spotted it or knew that it existed. Only a few theoretical glimmers predicted this faint glow from the early cosmos. The first came from physicist George Gamow, who was interested in the abundance of primordial helium in the universe. Helium is made up of two protons and a neutron or two. For the first few minutes after the big bang, when the universe was a sea of protons, neutrons, and electrons (and lots of photons), some of the protons and neutrons collided and made heavier elements, like helium, or deuterium (one proton and one neutron). Gamow realized that the universe's pressure, temperature, and density was related to how often protons and neutrons slammed into each other and formed elements heavier than hydrogen. Thus, he concluded, the amount of helium and other, heavier elements created in the first few minutes after the big bang should contain a wealth of information about the temperature, pressure, and density of the universe shortly after its birth. In 1948, the same year that saw the birth of the steady-state theory, physicists Ralph Alpher and Robert Herman used Gamow's idea to calculate the temperature of the radiation remnants from the big bang. The answer they got was a few degrees too high, but the important discovery was correct: there had to be a measurable amount of radiation left over if the universe had indeed been born in a big bang. However, their calculations didn't get all that much attention and sank into obscurity.

Nearly two decades later, Robert Dicke, an astronomer at Princeton famed for making sensitive antennas, came to the same conclusion through a different line of reasoning. According to Princeton astrophysicist P. J. E. Peebles, who was on Dicke's team in the 1960s, Dicke "did attend colloquia by Gamow, so he should have known about [the Alpher-Herman paper], but he felt he didn't." Ignorant of the work of Gamow, Alpher, and Herman, Dicke came to the idea of a cosmic background radiation because he was fond of the idea of an oscillating universe, one where it expands from a big bang, grows, and collapses again in a "big crunch," a big bang in reverse. From the ruins of collapsed universe, the next big bang ignites, and the cosmos is reborn, phoenixlike, starting the cycle over again.

However, to have a universe built from scratch each time, the heavier elements like uranium and oxygen and even helium have to be broken down to be recycled in the new universe. So Dicke did a few calculations. Just as an expanding universe cools down a contracting one heats up, and Dicke figured that the heating might break down the heavier elements and make them available for reuse. Even if the collapsing universe theory were incorrect, the calculations for determining the temperature of a universe just before it collapses in a big crunch would also apply in reverse; they would predict the temperature of the universe after the big bang. From those calculations, Dicke realized that there had to be a remnant radiation background left over from the big bang era. Dicke assigned two of his graduate students to build a microwave antenna to detect that radiation. Peebles refined the theoretical calculations, and like Alpher and Herman, he wound up with a too-high temperature. "I unknowingly reinvented Gamow's theory," says Peebles.

The stage was set for the discovery of the cosmic microwave background, but the Princeton scientists had their discovery snatched away from them. Two engineers at nearby Bell Laboratories, Arno Penzias and Robert Wilson, were trying to get rid of noise in their microwave antenna. At first, they thought it was caused by "a white material familiar to all city dwellers," but even after shooing the roosting pigeons out of the antenna and cleaning out all the droppings, the engineers could not get rid of stubborn microwave static coming from all directions in the sky. When Penzias and Wilson heard about the Princeton theory in 1965, they realized that the static was actually the afterglow of the big bang. It was the signal from shortly after the birth of the universe, a confirmation that the big bang was correct, and that the universe indeed had a beginning. For this, the two received the Nobel Prize.

The discovery of the cosmic microwave background was the last shot in the second cosmological revolution. In 1920, scientists had little idea whether the universe had a beginning or an ending; it was a question beyond the reach of science. When Hubble discovered that the universe was flying apart, he forced unwilling cosmologists to ponder how the universe was born and how it will die. Forty-five years later, when Penzias and Wilson discovered the cosmic background radiation in 1965, scientists had their first direct glimpse of the face of creation. The cosmic background radiation, the faint hiss of microwaves from all regions of the sky, is a snapshot of the infancy of our universe. It is the image of a moment a mere 400,000 years after the big bang, when all the matter in the cosmos was seething and glowing with the heat released from that monstrous cataclysm. No longer could cosmologists console themselves with the picture of an eternal, unchanging universe. All scientists now know the cosmos has a birth date, because we have seen the cosmos's baby pictures.

Chapter 4
The Third Revolution Begins
[THE UNIVERSE AMOK]

A coup or a palace takeover may be planned, but a revolution — never. Its outbreak, the hour of that outbreak, takes everyone, even those who have been striving for it, unawares. They stand amazed at the spontaneity that appears suddenly and destroys everything in its path. It demolishes so ruthlessly that in the end it may annihilate the ideals that called it into being.

— RYSZARD KAPUSCINSKI, *SHAH OF SHAHS*

Half a universe away, mismatched partners are locked in a dance of death. Two stars, near the end of their lifetimes, circle each other, held together by mutual, and deadly, gravitational attraction. One of the stars is shriveled, glowing white with heat. It is a white dwarf, the shrunken remnant of a star like our own sun, compressed into a space smaller than Earth. The other partner is bloated to enormous size; it is a cool, red monster, a giant, puffed out to many times its original girth and burning the last of its fuel.

The red giant and white dwarf orbit around each other, bound by each other's gravitational pull. Gravity, which created the pair, will also destroy them. Tugged by its partner, the swollen giant is pulled and distorted into a monstrous teardrop. Gas from the red giant streams from the teardrop's tip and spirals lazily into the white dwarf, like water running down a drain. As the white dwarf swallows the gas, month after month and year after year, it gets imperceptibly heavier. One day, all hell breaks loose.

When the white dwarf becomes too heavy—1.44 times the mass of our sun, to be precise—the extra mass it has accumulate crushes the uneasy equilibrium of the shrunken star. One dollop of gas too many and the star suddenly, catastrophically collapses. In a flash, the star crumples, heats up, and roars into a blinding explosion: a supernova. It is one of the most powerful events in the universe since the big bang, and it is a beacon visible throughout the cosmos.

The first two cosmological revolutions shattered the way we think about the universe and our place within it. The Copernican revolution destroyed the comfortable Aristotelian universe, where the Earth was safely ensconced within a tiny nutshell. Hubble's revolution and the discovery of the cosmic microwave background showed that the cosmos had a beginning and an ending. The supernovae are the harbingers of a third revolution. Scientists are now on the brink of answering eternal questions that have plagued humanity. Where did the universe come from? How will it end? Already, the revolution, being waged right now, has answered one of those questions, thanks to the stellar cataclysm half a universe away.

The third cosmological revolution took scientists by surprise, because they thought they already had a pretty good idea about how the universe worked. After the discovery of the cosmic background radiation, scientists understood the rough outlines of the birth of the universe. They knew that it was born in a flash of fire, and that the very fabric of space and time was expanding because of the cosmos-creating cataclysm. They didn't have many details. The exact age of the universe was known only as well as the rate of the expansion was— and measuring that expansion was very difficult, plagued with a great deal of uncertainty. Beyond that, cosmologists had little idea about the ultimate fate of the universe. They didn't know whether it would expand forever, or whether it would recollapse into a reverse big bang—a big crunch. These were important questions, but cosmologists thought they could answer them by taking more and more precise measurements. It might take decades or even longer to tie up these loose ends, but the universe seemed to have no big surprises left. Astronomers merely had to clean up the details.

In the 1980s and early 1990s, the story of the cosmos had a hazy beginning and no ending. In a slow, decades-long effort, scientists tried to make finer and finer measurements of the Hubble constant to try to clear up some of the haze; the more accurately they knew the speed of the expansion, the better they knew the age of the universe, and the better they would understand the birth of the cosmos. But at the end of the 1990s, those plodding, boring measurements suddenly turned very exciting. Instead of merely sharpening our existing measurements of the Hubble constant, supernovae from halfway across the universe changed the way we look at the universe, and they began to reveal not only how the universe was born, but how it will die. It was a stunning surprise, and it began a bewildered frenzy in the scientific world as experiment after experiment began to show that the universe is a much, much stranger place than we ever dreamed. In 1997, the third revolution in cosmology began. It is still under way.

The spark came from a fairly boring field. Supernova measurements were just a way to measure the expansion of the universe more accurately. It is a method not unlike what Hubble himself used. Like Hubble, the supernova hunters look for standard candles to help them gauge the distance to objects that are very far away. Hubble's favorite standard candles were the Cepheid variable stars. With the Cepheids he found in distant galaxies, Hubble not only showed that the universe was expanding, but also made a rough calculation of the current rate of expansion, a quantity symbolized by H_0, the Hubble constant. But getting a precise value is tricky. Hubble himself got it wrong, because astronomers of the day did not understand Cepheid variables terribly well.

Hubble's error came when he wrongly assumed that Cepheid variable stars all had the same properties. In 1952, astronomer Walter Baade showed that this assumption was wrong. Using the same hundred-inch telescope that Hubble used, Baade looked carefully at the Andromeda galaxy and proved that there are two types of Cepheid variable stars; one type (known as W Virginis stars) tends to be fainter than the other type. As a result, Hubble's standard candle wasn't quite as standard as he thought it was. This oversight led Hubble to conclude that galaxies were closer than they actually are. (It is as if you find out that a person standing next to an obelisk is about seven feet tall; you have to reestimate the distance to the obelisk, because you had made a slightly incorrect assumption about your standard ruler.) As a result, Hubble's value for H_0 was too large—he thought that the universe was expanding faster than it actually is.

This led to serious problems. The bigger the value of H_0, the faster the universe expands. But the faster the expansion of the universe, the shorter the time it takes to grow to its present size, so a fast expansion means that the universe is young. Conversely, a small value for H_0 means a slow expansion and an old universe. So, Hubble's too-large value of H_0 led to a too-short estimate for the age of the universe. Hubble's universe was only two billion years old, a figure that led to some glaring contradictions. For instance, by looking at the sizes, temperatures, and compositions of stars, stellar astronomers can figure out the stars' ages. Their calculations often showed that stars had been around for a lot longer than two billon years. It made no sense for stars to predate the universe they live in. Something had to be wrong.

Astronomers bickered for decades over the precise speed of the expansion of the universe. It was dreadfully embarrassing. The Hubble constant is a crucial weapon in the cosmologists' arsenal; it's hard to say that you have a grasp on the history of the cosmos if you can't even figure out its age. Hubble's discovery of the expansion was just the start of seven decades' worth of intense research. Scientists had to pin down the rate of the expansion by whatever means they could, and in the beginning of the 1990s, they had a powerful new tool to settle the argument: a space telescope, an orbiting observatory. The space telescope's name, Hubble, reflected its purpose. The Hubble Space Telescope's prime mission, the Key Project, was to determine, once and for all, how fast the universe is expanding.

Launched from the space shuttle *Discovery* in 1990, the Hubble Space Telescope is tiny compared to its earthbound companions. Its mirror is about ninety-five inches across, considerably smaller than the modern world-class ground telescopes. But it has one main advantage that ground-based telescopes don't have: it is above the atmosphere, so it has an incredibly sharp view of the heavens.

Though it seems transparent to us, the atmosphere is a roiling, semiopaque shield that blocks much of the light coming from the stars above. The light that does get through is distorted. Even on the clearest nights, the stars twinkle — they fade in and out, and even seem to dance a little bit. This happens because the atmosphere is such a turbulent place. Even though casual observers don't notice the turbulence except as a faint twinkling of the stars, it frustrates earthbound astronomers, who are trying to discern very fine details in distant galaxies and nebulae. The constant shimmering of the atmosphere ruins their pictures, blurring them and destroying those details. Though astronomers now have ways of reversing some of the effects of the roiling atmosphere—with flexible mirrors and "adaptive optics" they can cancel out some of the motion—the best of all possible options is to get outside the atmosphere altogether. An orbiting observatory also has another advantage; it can see colors of light that are blocked by the atmosphere (like ultraviolet light, x-rays, and gamma rays) or overwhelmed by earthly sources of radiation (like infrared rays or microwaves). This explains why astronomers can convince politicians to spend so much money on space telescopes.

Once NASA corrected the Hubble Space Telescope's early problems (the mirror manufacturer goofed, forcing NASA to refocus the telescope in 1993), it became a spectacular tool for astronomers, observing the heavens by detecting infrared rays and ultraviolet light, as well as the everyday optical spectrum of the rainbow. Astronomers working with the Hubble telescope made lots of pretty pictures of exotica in space, but more important, they gathered reams and reams of data on Cepheid variable stars in an attempt to calculate the Hubble constant. They also used other methods to try to pin down the Hubble constant, such as measuring spiral galaxies' rotation rates, which are related to their intrinsic brightness. (This relationship between rotation and brightness, called the Tully-Fisher relation, was discovered in the 1970s and turned spiral galaxies into standard candles. Though their brightness is known less accurately than that of Cepheid variable stars, they are much brighter, so they are visible from much farther away.)

In 1999, after six years of observations and analysis, the Hubble Key Project was complete. The scientific team, led by Wendy Freedman of the Carnegie Observatories in Pasadena, released its value for the Hubble constant: 72 km/sec/Mpc.[1]

---

1. The unit of the Hubble constant is a wee bit complicated, but it makes sense if you think about it. H_0 measures a relationship between speed and distance: the farther away an object is, the faster it is receding. The km/sec part is the speed; the Mpc

And even all that work didn't end the controversy. "That's a bunch of hooey," said Allan Sandage, also of the Carnegie Observatories, who also used data from the Hubble Space Telescope. Sandage came up with a value of H_0 closer to 60, considerably slower than Freedman's value. Even the billion-dollar space telescope didn't settle the issue. Luckily, lots of other astronomers were trying to figure out the value of the expansion as well, using other techniques. One of the most promising had to do with supernovae, the cataclysmic death throes of massive stars.

When a star is massive enough—above the *Chandrasekhar limit* of 1.44 times the size of our sun—its death is violent and spectacular. Throughout its normal lifetime, a star is an uneasy battleground between the force of gravity and the energy of nuclear fusion. Gravity tries to compress the star into a tiny ball, while the heat and light of its nuclear furnace (mostly turning hydrogen into helium) tries to blow it apart. But after millions or billions of years, a star begins to run out of hydrogen to burn. It begins to fuse helium and then heavier and heavier elements in a desperate attempt to keep itself from collapsing. Eventually, however, it runs out of fuel.[2] The nuclear furnace can no longer counteract the pressure of gravity, and gravity finally overcomes the internal pressure of the star. The star collapses, releasing a flash of energy.

A relatively light star, like our own sun, will release a moderate amount of energy as it collapses to form a white dwarf, a midget star roughly the size of the Earth. What keeps it from collapsing entirely is that the star's electrons re-

---

(megaparsec, a unit of distance roughly equivalent to 3.26 million light-years) is the distance. So, for every megaparsec more distant from Earth you go, the average galaxy will be speeding away about 72 kilometers per second faster. For the sake of clarity, km/sec/Mpc will henceforth be omitted.

2. Iron is the most stable nucleus, so, in a sense, every element "wants" to be iron. Taking hydrogen atoms and fusing them into helium gets the ensemble closer on the periodic table to iron, so the reaction releases energy. This works for elements lighter than iron, but as soon as you try to fuse iron with something, the atoms are moving away from iron on the periodic table so the reaction doesn't release energy; it absorbs energy instead. This is what makes the nuclear furnace run out of fuel.

pel one another: when gravity tries to force them to occupy the same spot, they resist, counteracting the force. But even electron repulsion has its limits. If the star is larger than the Chandrasekhar limit, the mass of the star is so great, and its gravitational pull is so strong, that it overcomes the electron repulsion and collapses even further, becoming a neutron star or a black hole. When it does this, it releases an enormous amount of energy. It dies in a supernova. Supernovae are the most energetic explosions in the modern universe, and they are visible for billions of light years in all directions. And, luckily for scientists, one type of supernova, known as a type Ia, is a standard candle that, unlike the Cepheid variables, can be spotted half a universe away.

Other supernovae begin with stars of different masses, so when they explode, they release wildly different amounts of energy. A type Ia, however, explodes in the same way each time, because of its peculiar history. It is the product of a dance of death between two stars, a greedy white dwarf and its swollen partner. Over time, the dwarf gets heavier and heavier, as it steals more and more gas from its hapless companion. At the moment that the dwarf finally surpasses the Chandrasekhar limit, it explodes in a supernova. That is, a type Ia goes off exactly when a sub-Chandrasekhar-limit white dwarf happens to cross the fatal threshold and go supernova. Each of these stars is the same weight when it explodes— at exactly the Chandrasekhar limit—and so it blows up in much same way each time, with the same mass, and the same energy. And the same brightness. It is a standard candle.

Two competing teams of astronomers have been spending years studying type Ia supernovae, hoping to use them to figure out the Hubble constant. The Supernova Cosmology Project and the High-Z Supernova Search Team[3] used data from the Hubble Space Telescope, the Cerro Tololo Inter-

3. Z is the term astronomers use to measure extremely great distances; it is related to the redshift of objects. The higher the Z, the greater the redshift, and the more distant an object is, thanks to the Hubble relationship between redshift and distance.

American Observatory in Chile, the Keck telescopes in Hawaii, and a number of other telescopes across the world to find and measure type Ia supernovae. The goal was to measure the rate of the Hubble expansion of the universe—now and in the past. Because supernovae are visible at such great distances—the current record is more than ten billion light-years away—supernovae allow astronomers to calculate the Hubble expansion not only in the present-day universe, but in the past as well. When astronomers look at a supernova a billion light-years away, they are looking at light that left the supernova a billion years ago. When scientists gaze across great expanses of space, they are actually looking backward in time. This is valuable because the Hubble constant had to change as the universe evolved.[4]

The universe is expanding because it got an enormous kick when it was born. It is like hitting a baseball with a bat—one good smack sends it flying vertically into the air. But as time passes, the baseball's speed drops more and more as gravity pulls on it. At some point, the baseball's vertical motion stops entirely, and it drops back to the ground. Some cosmologists thought that the universe might work something like that; after the original kick, the universe's expansion slows and slows, and eventually stops. After that, it would begin to collapse under its own gravitational pull, like the baseball falling back to the ground. The universe would shrink and shrink, heating up and eventually disappearing in a big crunch—a big bang in reverse. However, cosmologists saw another possibility. In theory, you can smack the ball so hard that it escapes the earth's gravitational field; the ball slows down, but it retains enough of its original kick to fly out into space. The ball never falls back to Earth. It would get slower and slower and slower, but it would fly into deep space and out of the solar system. It would never return to Earth. Some cosmologists believed that the universe behaved like this, always expand-

4. Constants don't usually change, but to astronomers, the "constant" simply refers to the present-day expansion of the universe. More detail follows shortly.

ing, losing speed as it went, but never collapsing upon itself. The universe would grow larger and larger, cooling and eventually dying out as the stars burned up their last bits of fuel.

No matter whether they believed in a big crunch or an ever-expanding universe, all cosmologists agreed that the Hubble constant, like the speed of the baseball, was not so constant after all. It had to be greater in the past than it is today, just as a baseball had to be moving faster when it was hit than when it had already traveled in the air for a few seconds. The mathematical symbol for the Hubble constant, H_0, reflects this changing nature of the expansion of the universe; the subscript 0 denotes the present rate of expansion. The Hubble constant should be smaller now than it was billions of years ago, just as a baseball high in the air has to be moving slower than it was right after it was hit.

The two teams searching for type Ia supernovae did not expect to see anything unusual when they analyzed their data—the supernovae that they meticulously collected from telescopes across the world. For them, each supernova is like a snapshot of the past, and a way to figure out the speed of the expansion in the ancient universe. By collecting enough supernovae at various distances, both teams hoped to assemble a picture of the Hubble constant at different times in the past and to measure how the expansion of the universe was slowing down over time. However, in 1997, the two teams finally had the data they needed, and they gave the first glimmers of an answer to one of the oldest questions known to humanity: how will the universe end? In the process, the teams inadvertently opened up the most bizarre, disturbing, and exciting chapter in cosmological history.

Each supernova is a single data point, a measurement of the Hubble expansion at one moment in the past. Both teams needed enough of these points to create a coherent picture of the history of the expansion, enough snapshots to cobble together a movie that describes the evolution of the universe over the course of billions of years. In late 1997, Saul Perlmutter of the Supernova Cosmology Project and Brian Schmidt of the High-Z Supernova Search Team had enough supernovae to assemble the first crude history of the universe's expansion throughout time.

Ancient myths often told tales of how the universe is destroyed, but modern science has been grappling seriously with the question only since the second cosmological revolution. The work of Hubble forced cosmologists to confront the beginning and end of the universe, but for decades, they did not have the tools to determine the ultimate fate of the cosmos. They debated about how the universe would end: eternal expansion or big crunch? Einstein's equations imply that one or the other must be true; the universe is an ever evolving battle between gravity and the force of the big bang, and nobody knew which side would win. The victor would decide how the cosmos died. The supernova teams, in measuring how rapidly the expansion of the universe was changing, were trying to figure out nothing less than the ultimate fate of the cosmos.

If gravity wins out in the end, if there is enough matter in the universe to overcome the expansion caused by the big bang, then the expansion will slow and slow and slow, eventually halting entirely. In this scenario, as astronomers observe the universe over billions of years, they see the Hubble constant decreasing and decreasing, dropping to zero. The expansion of the universe has been entirely negated by the gravity of galaxies. But the story does not end there. Gravity continues pulling things together even after the expansion peters out, and the universe begins to collapse. It gets smaller and smaller and smaller, faster and faster and faster. The collapse of the universe accelerates, like a baseball falling to Earth. Just as the universe cooled as it expanded, it heats up as it collapses. Soon, everything is bathed in a fiery glow of light: ultraviolet, then x-rays, then gamma rays. Even matter itself is no longer able to survive in the bath of radiation. Electrons fly away from their atoms, and even the nuclei of atoms fly apart. In the last moments of the dying cosmos, the very protons and neutrons that make up the nuclei disintegrate, and the universe is swallowed up into an anti-big-bang: the big crunch. The universe dies a death by fire.

On the other hand, theorists realized that there might not be enough matter in the universe to counteract the force of the initial explosion. Though the expansion of the universe slows and slows—observers see the Hubble constant dropping off over time—the expansion never stops entirely. Galaxies fly away from one another forever, getting farther and farther away, dimming and reddening over time, disappearing from the skies. Stars would burn out and die, becoming empty shells of cold, dead matter. The stars would flicker out, one by one. The universe would get cooler and cooler over time, and the last bits of matter might even disintegrate, decaying into energy and giving brief bursts of light to the cold soup of lifeless radiation that suffuses the universe. There would be nothing remaining except for a frigid bath of cold light. It would be a death by ice.

Hunting for supernovae is difficult and time-consuming. Astronomers take pictures of vast sections of the heavens, over and over and over again. Once in a while, there is a tiny change—in one patch of the sky, there's a bright spot that wasn't there before. If it is a type Ia supernova, they estimate its distance by looking at its brightness, and compare that to its redshift. Over time, the supernova hunters built up a database of supernovae at various distances from Earth. By comparing the distances of these supernovae with their redshifts— the rates at which they were speeding away—the supernova hunters could see how the rate of the universe's expansion slowed over time. If the universe was slowing down rapidly, the universe would collapse in a fireball; if it was decelerating very slowly, the universe would cool and expand forever.

When the two teams released the first supernova results, they showed that the expansion of the universe was not slowing down very much. Perlmutter, Schmidt, and their colleagues revealed for the first time that the force of gravity was losing the battle, and the universe was expanding unabated. Most scientists now believe the expansion of the universe will continue forever. Our destiny is a death by ice.

It was a stunning discovery. For the first time in history, scientists had wrested the end of the universe from the hands of mythology and speculation and placed it firmly within the grasp of human knowledge. This will be one of the most enduring victories of cosmology. However, this was just the beginning of the third cosmological revolution, because this understanding, the knowledge of the fate of the cosmos, came at a grievous price. It upset scientists' very conception of the nature of the universe.

In 1998, less than a year after the initial announcement of the fate of the cosmos, after the supernova teams had gathered more data and had done some more calculations, the story got considerably weirder. The supernova hunters had continued to measure the rate of expansion of the universe, now and in the past, but with their additional data they saw something utterly bewildering. In 1997, they had realized that the expansion of the universe wasn't slowing very much. In 1998, however, they saw that the expansion of the universe wasn't slowing at all; in fact, it was speeding up. This was as unexpected as seeing a baseball that was hit into the air rushing faster and faster upward, ever accelerating. It was as if they had spotted some sort of bizarre antigravity force, propelling the baseball into the air at ever increasing speeds. Cosmologists were dumbfounded by the revelation.

All the models of the universe had assumed that the universe's expansion was slowing down over time. The supernova hunters discovered that the Hubble constant—the speed of the expansion—was smaller in the past than it is now. A commonsense assumption was utterly wrong. Cosmologists had to throw out one of the most basic ideas that they had learned in school, and in an ironic twist, they began to explore Einstein's "biggest blunder." The greatest mistake of Einstein's career might not have been such a mistake after all.

The cosmological constant, Λ, kept the universe from collapsing—it was a mysterious repulsive force that Einstein artificially inserted into his equations and then discarded when Hubble discovered that the universe was, indeed, expanding. The supernova data forced cosmologists to take a second look at Λ, nearly seventy years after its creation. Since the universe is expanding faster and faster, rather than slowing down as everybody expected, astrophysicists had to consider the possibility of some sort of antigravity force, some mysterious essence that resists the pull of gravity. The ever accelerating expansion of the universe points to the existence of a repulsive force: something is inflating the balloon universe more and more vigorously. Nobody really knows what it is, though they now have a few theories. Nevertheless, this repulsive force, Λ, instantly became the biggest mystery in science.

Chapter 5
The Music of the Spheres

[THE COSMIC MICROWAVE BACKGROUND]

Our impotence to hear this harmony seems to be a conse-
quence of the insolence of the robber, Prometheus, which
brought so many evils upon men. . . . If our hearts were
as pure, as chaste, as snowy as Pythagoras' was, our ears
would resound and be filled with that supremely lovely
music of the wheeling stars. Then indeed all things would
seem to return to the age of gold.

—JOHN MILTON, "ON THE MUSIC
OF THE SPHERES"

The supernova hunters left cosmology in disarray. While they charted the ultimate fate of the universe, they stumbled across the influence of a mysterious force—a cosmological constant—that was blowing the universe apart, making it expand faster and faster. Naturally, scientists were reluctant to accept the idea of an unknown antigravity force, but the data were forcing them to accept the ridiculous picture of the universe that went along with it. They began looking for a way out. Some argued that the supernovae aren't quite the standard candles that people assume. If, for instance, super-novae were fainter in the past than they are today, they might seem more distant than they actually are, throwing off the calculations.

Proving the existence of something as counterintuitive as the cosmological constant takes a great deal of evidence. The supernova data were compelling hints of an accelerating universe, but if they had been the only indication of the anti-gravity force, scientists probably would have written them off as an interesting experimental quirk. But there are other, more powerful methods of divining the fate of the universe, and they too are finally coming into the reach of astronomers and are confirming the increasingly bizarre picture of the universe. Scientists are learning to read the writing on the walls of fire that surround us.

The cage of fire, the cosmic background radiation, is inscribed with the history of the infant universe. It tells not only of how the universe was born, but also of what it contains and how it will end. In the spring of 2000, scientists were finally able to read the writing hidden within the cosmic background radiation. A balloon-borne experiment, Boomerang, returned the first high-resolution map of the small-scale patterns in the cosmic microwave background, and that was just the beginning. A year later, three cosmic-background teams released their data simultaneously, jointly providing the finest picture ever of sections of the universe's edge. In 2002 a satellite orbiting Earth was doing the same for the entire sky. These new measurements promise to reveal the origin and the fate of the universe in a way that we could scarcely have dreamt of a mere decade ago.

Cosmologists have begun to decode what the cosmic background radiation is telling us. They are already astonished at what they have read.

❊   ❊   ❊

To understand the cosmic background radiation, we must travel back to the very beginning of the universe, to the big bang itself. At first glance, the history of the universe, as told by a scientist, will seem almost as farfetched as the stories told by Greek mystics or African storytellers. However, unlike the ancient myths, every part of the scientific narrative is backed up by hard scientific evidence, evidence that will become clear as we get deeper and deeper into the story. And as strange as this tale may seem, scientists were forced to accept it to explain their observations of the heavens.

Like many ancient myths, the beginning of the universe, as seen by modern science, begins with nothing at all. There is no space; there is no time. There is not even a void. There is nothing.

In an instant, the nothing becomes something. In an enormous flash of energy, the big bang creates space and time. Nobody knows where this energy came from—perhaps it was just a random event, or perhaps it was one of many similar big bangs. But within a tiny seed of matter and energy is all the stuff of our current universe. For a fraction of a second the universe expands at an incredible rate; it is inflated by an energy that scientists do not understand very well. But that brief era of inflation, as chapter 12 will explain, leaves its mark upon the face of the modern cosmos.

As the newborn universe expands, matter begins to coalesce. About a trillionth of a second after the big bang, quarks, gluons, and leptons begin to form out of the radiation. These are the fundamental building blocks of matter. We will encounter these particles in more detail a little later, but a brief introduction is in order. Quarks are relatively heavy, indivisible chunks of matter that make up the stuff at the center of atoms, and gluons are "sticky" particles that make quarks cleave together. Leptons, like the electron, are lighter indivisible particles. Unlike quarks, leptons don't feel the pull of gluons. (It might seem strange that some particles feel the pull of gluons and others don't, but it is an ordinary occurrence even in the macroscopic world: paper clips feel the pull of a magnet, but pennies don't.) At this point, it's not crucial to delve into the details; all you need to know is that quarks, gluons, and leptons are the most primitive matter in the universe, and until about a millionth of a second after the big bang, the universe is a seething soup of primitive matter and radiation. Quarks, gluons, and leptons float about, battered by radiation, in what physicists call a *quark-gluon plasma.*

After about a millionth of a second has gone by, the soup cools to ten trillion degrees. The particles of matter lose energy and slow down. The quarks can no longer resist the attraction of the gluons, and they begin to succumb to the pull. The quark-gluon plasma begins to coalesce, much as a cloud of water vapor condenses into droplets when it encounters a cool windowpane. When they stick together, quarks condense into particles like protons and neutrons, the heavy particles at the center of atoms, as well as more exotic particles that do not survive as long.

As the universe cools, all the quarks in the universe condense into particles like protons and neutrons. These newborn protons and neutrons are still very hot—they have a lot of energy—so they fly around erratically, colliding into one another, sticking together and splitting apart. However, as the universe cools and expands yet more, the protons and neutrons slow down. But just as quarks stick together once they cool below a certain temperature, so do protons and neutrons. When the protons and neutrons slow down enough, they no longer have the energy to split apart again when they collide. Protons and neutrons coalesce, forming atomic nuclei: deuterium, helium-3, helium-4, and other heavier species. This is the era of *nucleosynthesis,* and it was incomplete. Most of the protons in the universe remained unbound, solo, protons, and a lone proton is the core of what chemists call hydrogen.

Almost all the matter in the universe was born in the first

THE UNIVERSE SO FAR

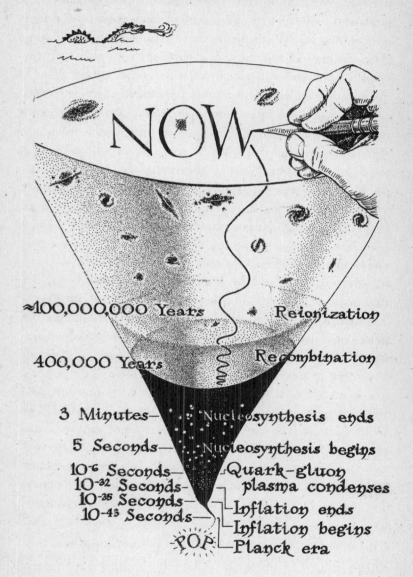

≈100,000,000 Years — Reionization

400,000 Years — Recombination

3 Minutes— Nucleosynthesis ends

5 Seconds— Nucleosynthesis begins

10^{-6} Seconds— Quark-gluon
10^{-32} Seconds— plasma condenses
10^{-35} Seconds— Inflation ends
10^{-43} Seconds— Inflation begins
POP— Planck era

few minutes of creation. Quarks and leptons were created out of energy, then the quarks coalesced into protons and neutrons, and then the protons and neutrons wound up forming light atomic nuclei like deuterium, helium-3, and helium-4. (This last process is very similar to what happens in the center of the sun and at the heart of a hydrogen bomb.) After a few minutes, the hot, dense, pressure-cooker universe cooled so much that heavy nuclei could no longer form. Protons and neutrons did not slam into each other with sufficient energy to make them stick together. A few minutes after the big bang, the process of nucleosynthesis stopped. Hydrogen nuclei stay hydrogen nuclei and deuterium nuclei stay deuterium nuclei. Almost all the matter in the universe, such as the protons and electrons that make up the hydrogen atoms in a glass of tap water, is exactly the same as it was when it was formed in the first moments after the big bang.[1]

After the universe is too cool for nucleosynthesis, the quantities of hydrogen, deuterium, helium-3, helium-4, and the other primordial chemical elements are frozen—to use the term very loosely, because the universe is thousands of degrees Celsius. This is too hot for ordinary atoms, the sort of matter we encounter every day, to form.

An atom has two parts. Its nucleus is made of protons and neutrons (except for hydrogen, which has no neutrons). The nucleus is relatively massive, because protons and neutrons are heavy particles. A cloud of electrons, very light elementary particles, surrounds that nucleus. The electrons are bound to the nucleus by electromagnetic attraction just as the moon is bound to the Earth by the force of gravity. However, in the

---

1. Heavier elements, like oxygen and carbon, are extremely rare in the universe, even though they are relatively abundant on Earth. The big bang didn't make these heavier elements; they were formed in the heart of a star, which, like the pressure-cooker early universe, was hot and dense enough to make nuclei fuse together. All the heavy elements on Earth were made in a star that exploded, scattering the seeds of life in an expanding cloud. This is why Carl Sagan often said that we are all made of starstuff.

hot early universe, there was too much energy available for electrons to bind to their nuclei. Radiation, extremely energetic light, was everywhere. These photons, these particles of light, prevented electrons from binding with nuclei. The early universe glowed brightly. Photons skittered in all directions. Every time an electron settled down with a nucleus, a photon hit it and sent it flying. The universe had to cool down, and the barrage of photons had to become less energetic and less dense before electrons could settle down, bind to nuclei, and form atoms.

We have seen this sort of condensation process twice before. About a millionth of a second after the big bang, the universe cooled enough for free quarks to bind together, forming protons and neutrons out of quark-gluon plasma. A few seconds later, the protons and neutrons had cooled enough so that *they* could bind together, and for several minutes they condensed into various nuclei during the era of nucleosynthesis. Similiarly, the universe eventually cooled enough so that nuclei and electrons could finally bind together and form atoms, the process known as recombination. But this third time, it took the universe much, much longer to cool and condense. It took about 400,000 years to get to the three thousand degrees when recombination occurs.

In the intervening time, from few minutes after the big bang until recombination 400,000 years later, the nuclei of atoms—the protons and neutrons—were not tied to the light, negatively charged electrons. In such a state, matter behaved very differently from the way it behaves in the solids, liquids, and gases that we encounter every day. In ordinary matter, any given electron tends to be tied to a specific atom; occasionally, one gets knocked free, but it quickly settles back down and attaches itself to another nucleus. But when matter gets very, very hot, as in the early universe, the electrons form a polygamous soup with their nuclei. No electron is tied to any one nucleus. This makes the energetic matter very different from a solid, a liquid, or a gas. Scientists call this state a plasma.[2]

Plasmas have certain interesting properties: they conduct electricity the way a metal does, and like metals they are opaque to light. (In fact, the structure of metals allows nuclei to share electrons, which is why metals and plasmas have some of the same properties.) Any photon venturing into a plasma will not get far. It will quickly be scattered or absorbed by the electrons and nuclei.[3] So, before recombination, before the time when the electrons settle down with their nuclei, it was impossible for a photon to get very far without hitting an electron. Every photon scattered and rescattered; the universe was opaque to radiation and did not let photons travel freely. Light was trapped in a cage of matter, and the whole universe was a seething, opaque mass that cooled for thousands of years.

All of a sudden, 400,000 years after the big bang, recombination hit. After spending years and years living together, the electrons and nuclei decided to settle down together. As the electrons bound with nuclei, forming atoms, the photons were released from their cage of matter. The opaque universe suddenly became clear, and after their last scatter, the photons were free to zoom off in every direction.

When Penzias and Wilson detected the cosmic background radiation, the mysterious "static" in every direction of the sky, they were really seeing a faint image of the *last scattering surface,* the cloud of plasma that scattered the light for the last time before setting it free. The static in the sky was really the radiation released during recombination, stretched and attenuated over fourteen billion years. This cosmic microwave background is the most ancient light astronomers will ever

2. The quark-gluon plasma was named by analogy with the plain-vanilla plasma; in a plasma, electrons and nuclei roam free, just as quarks and gluons roam free in a quark-gluon plasma.
3. The U.S. and Russian militaries have experimented with creating clouds of plasma around airplanes, because the plasma will absorb incoming radar beams, which are, after all, just light rays.

see. It surrounds us in all directions; we are trapped within walls of fire. The background radiation is the image of the fiery walls of the universe.

The cosmic microwave background gave scientists their first direct glimpse of the immediate aftermath of the big bang, and it confirmed big bang theory and killed off steady-state theory. Though the other stages in the story of the big bang will become important later in this book, the era of recombination has become crucial to our understanding of the origins of the cosmos. The fiery walls of the universe are inscribed with clues about the beginning of the universe. But for years, scientists could not read those clues.

The problem was that the cosmic microwave background is a faint whisper, a quiet echo of the incredible blaze of light that escaped from its cage of matter. Starting off as ultra-high-energy gamma rays, the light was stretched and stretched, losing energy all the while. It quickly dimmed and cooled, turning into x-rays, then ultraviolet light, then visible light, finally passing into the infrared and microwave regions of the spectrum. Microwaves are very hard for an earthly observer to pick up, because everything on the planet, telescopes included, radiate microwaves, swamping the signal from fourteen billion light-years away.[4]

Every object has a temperature. In a sense, temperature is a measure of how the atoms in an object are vibrating. And everything that has a temperature radiates light. When commandos put on their night-vision goggles, they are trying to pick up the light radiated by warm objects like people. Living creatures tend to radiate strongly in the infrared region of the spectrum, which is what night-vision goggles are tuned to de-

4. Unfortunately, the combined nature of space and time makes it rather confusing when describing something that happened very long ago or very far away. Recombination happened a little less than fourteen billion years ago, so in a sense it happened a long time in the past. But we are just getting the light waves from that era that have taken fourteen billion years to get here, so in another sense it is a very real event that is happening right now, but very far away. When we look at objects very far away, we are also looking backward in time.

tect. Hotter objects radiate more light energy. When magma glows red or an iron bar glows white-hot, you are detecting the higher-wavelength, visible, light that comes off those hot objects; they are so hot that you don't need night-vision goggles to see them in the dark. Conversely, cooler objects, like an ice cube or a tub of liquid nitrogen, radiate less energy— almost no visible light and little infrared energy. However, they still radiate microwaves, which have less energy than even infrared light.

The temperature of an object dictates the amount and kind of light that a generic object gives off; this relationship is encoded by the *blackbody* spectrum. A blackbody is an ideal object, something that does not reflect any light, but instead absorbs it, converts it into thermal energy, and emits it as light. (The name is a little confusing. Blackbodies aren't necessarily black; if they're hot enough, they can shine with white light.) Blackbodies emit energy based only upon temperature; if you measure the light that comes from a blackbody object, you can figure out what temperature it is. The last scattering surface, the wall of plasma that generated the cosmic background radiation, behaved pretty much like a blackbody. After billions of years of stretching, though, the blackbody radiation looked as if it were coming from an object that was only 2.7 degrees above absolute zero. Almost everything in the universe—glowing iron rods, people, ice cubes, and even the Earth itself—shines with enough microwaves to overwhelm the signal.

As a result, to see the cosmic microwave background, scientists have to block out all the competing signals from surrounding objects. It is such a tricky measurement that the first decades of cosmic-background measurements provided very little information other than the mere existence of the radiation at 2.7 degrees. Scientists couldn't even prove that the background radiation had a blackbody spectrum until they had been hard at work for a quarter of a century. Their detectors simply weren't good enough to take measurements with the needed precision. However, even though the experimentalists were stymied, theorists raced ahead with their predictions about what the cosmic microwave background would tell us if we ever got sufficiently sensitive instruments. (Lack of experimental information never stopped a good theorist.) It turns out that the cosmic microwave background has plenty to say.

For one thing, the infant cosmos was not a uniform cloud of plasma. In some places the glowing plasma was thicker and denser, and in others the cloud was thinner and less dense. As the universe expanded, the matter in the thick, dense parts collapsed under its own gravity, forming galaxies and galaxy clusters. The thinner regions got more and more sparse as the universe expanded, forming bubblelike voids between galaxy clusters. The thick and thin regions in the plasma—the *mass fluctuations* in the primordial universe—should have left their mark on the cosmic background radiation. This means that the faint microwave light from the edge of space should contain information about the matter and energy that filled the universe.

In the early 1970s, Yakov Zel'dovich and other physicists investigated the cosmic background radiation. Perhaps their most important realization was that the light and matter in the pre-recombination universe was seething and oscillating, which should have made the cosmic background radiation lumpy rather than smooth. Instead of a uniform hiss of noise coming from everywhere in the sky, the mass fluctuations have produced patches where the hiss is stronger and "hotter," and other patches where the hiss is weaker and "cooler." The hot zones are where the plasma cloud was particularly dense, and the cool parts are places where the plasma was sparse. This patchiness in the cosmic background radiation is called *anisotropy*, thanks to the astronomers' love of Greek words.[5] By analyzing the nature of that anisotropy, scientists

5. *Isotropic* refers to something that is the same in all directions; *iso* is a Greek prefix that means "same," and *tropos* means "disposition" or "character." Anisotropic is the opposite; it refers to something that is not the same in every direction.

could, in theory, figure out what sort of matter and energy were rattling around in the early cosmos. But until 1990, cosmologists failed to see any evidence of anisotropy; their instruments simply weren't good enough. With blurry glasses, they couldn't see any writing on the walls of the universe.

In 1990, however, a satellite known as the Cosmic Background Explorer (COBE) orbited far above the Earth. Sitting in the void of space, chilled with liquid helium, and shielded from radiation from the Earth and the sun, COBE detected anisotropy in the cosmic background radiation for the first time, though it could not make out the fine details. It was something like hearing the faint thrum of music a long distance away but being unable to name the tune. This anisotropy was very slight, though. Imagine that you are in a crowd of six-foot-tall people. If the cosmic microwave background were isotropic, it would be as if they were all exactly the same height, whereas an anisotropy would show up as a slight variation in heights from person to person. The anisotropy that COBE detected in the cosmic background radiation was the equivalent of a large crowd of six-foot-tall people differing in height by about a quarter of the thickness of a human hair. As tiny as this anisotropy was, COBE proved that it was there. (COBE also proved that the cosmic microwave background looked as if it had been emitted by a blackbody.) But COBE, as well as a lesser-known balloon experiment that went up at the same time, could not do anything more than detect the anisotropy. It could not make out the fine details. COBE had proved that the walls of the universe were covered with writing, but it wasn't able to read them. Cosmologists were desperate to understand the secrets hidden within the writing.

The ancient Greek philosophers thought that the universe was filled with music. Humans cannot hear the music of the spheres, they argued, because our world is made of different stuff from that of the heavens above. In the minds of the ascetic Christian thinkers, our mortal world is tainted by the impurity of original sin. Our mortal senses are unable to discern the empyrean harmony. The blind John Milton said that if we could hear the celestial music of the spheres, time would reverse itself, and we would glimpse the purity of Eden and the golden age of the universe. In a sense, Milton was right.

Zel'dovich and other physicists realized that the early universe was ringing like a bell, and for years cosmologists tried in vain to listen to that ancient celestial music. The cosmic background radiation is an echo from an age when the entire universe was an enormous musical instrument, ringing with the sound from the big bang. When we finally sharpened our senses to where we could hear that celestial music, it transported us back in time to the infancy of the universe.

When you hear the faint tintinnabulation of a church bell in the distance, your eardrums are picking up tiny fluctuations in air pressure. When a clapper strikes a bell, the metal of the bell vibrates in a specific way, depending on the size and shape of the bell, the material it is made of, and a variety of other factors; these characteristics determine what note the bell will sound. The vibrating bell, in turn, causes the air around it to slosh back and forth, causing *pressure waves*. When the bell pushes a clump of air, it compresses it, making it somewhat denser than the surrounding air. But the molecules of air bump into one another more often when they are compressed, forcing one another apart. Within a fraction of a second, the compression—the *overpressure*—blows itself apart and the clump becomes less dense than the surrounding air, becoming an *underpressure*. Air rushes in to fill the underpressure, becoming overpressure again, and so forth—the bell makes the air oscillate between overpressure and underpressure. Your eardrums are pushed inward by overpressure and pulled outward by underpressure, and your brain turns the eardrum oscillations into the sound you perceive. For instance, when the air makes your eardrum flap back and forth roughly 262 times a second, you perceive a middle C; the faster the flapping, the higher the note you hear.

But there is more to an instrument than its pitch. A violin, a clarinet, and the human voice can all produce the same middle C, but your ear can distinguish among the three sounds. That is because these instruments don't produce pure tones. When a violin plays a middle C, the air oscillates at other frequencies besides 262 times a second. (A pure tone sounds more like a tuning fork or a computer beep than a musical instrument.) The violin is specifically constructed to vibrate in complicated ways. A violin's middle C contains not only the 262 Hz *fundamental* frequency, but lots of higher-frequency overtones: 524 Hz, 786 Hz, 1048 Hz, and innumerable others.[6] The relative strengths of all these overtones and the impurities in them are largely responsible for the difference between the syrupy wail of the cello and the tinny whistle of a flute, or the bass rumble of a distant drum. The various oscillations that strike your eardrums encode the information about the instrument that sounds the note.

Similarly, oscillations of the early universe encode information about the nature of the cosmos in its infancy. In fact, those oscillations are very much like sound waves, which is why scientists call them acoustic oscillations. However, those oscillations are on a much grander scale than the air waves that are used to. The entire universe is the greatest instrument ever known, and the cosmic microwave background is its distant echo.

Just as sound waves are alternating compressions and rarefactions of air, the acoustic waves in the early universe are caused by alternating compressions and rarefactions of plasma in the pre-recombination universe. These waves were created because matter in the primordial plasma was caught between two competing forces. For one thing, there was

6. A hertz, Hz, is the scientist's shorthand for one action, such as an eardrum flap or a metronome tick or a wheel's revolution, per second.

gravity. Gravity, though it is a relatively weak force, tries to pull matter into clumps. If gravity were the only force in the universe, every bit of matter in the universe would be stuck in one gigantic lump. Luckily for us, other forces battle the effects of gravity. In the early universe, the important one was *radiation pressure*—the force exerted by photons on the plasma.

Recall that the electrons and nuclei in a plasma float free, and a photon cannot penetrate very far in a plasma before it scatters off a particle. Looking at it from an electron's point of view, it's a fairly uncomfortable place. The electron is under a constant barrage of photons, getting battered left and right by particles of light. When an electron is hit by a photon, it gets a kick of energy that sends it zooming off in the opposite direction. In essence, the photon pushes the electron away—the radiation exerts pressure upon matter. Just as gravity tries to pull particles together, the intense light in the plasma drives the particles away from each other. The hotter a clump of matter gets, the more photons it emits, and the stronger the radiation pressure gets, driving the clump apart more vigorously.

The two opposing forces of gravity and radiation pressure caused the primordial plasma to oscillate in the following way. Imagine a small cloud of primordial plasma in the early universe. As gravity gets the upper hand in this clump of matter, the clump begins to compress and heat up.[7] The clump emits more light, and the radiation pressure increases. Pretty soon the outward push of the light matches the inward pull of gravity, and then the outward force overmatches gravity, forcing the clump to expand. The expanding cloud cools and produces less light. The radiation pressure goes down, and the force of gravity gets the upper hand once more. Starting in the very early moments of the universe, clumps of matter oscillated, compressing and expanding, caught between the two competing forces.

7. Remember, things heat up as they get squished and cool down as they expand.

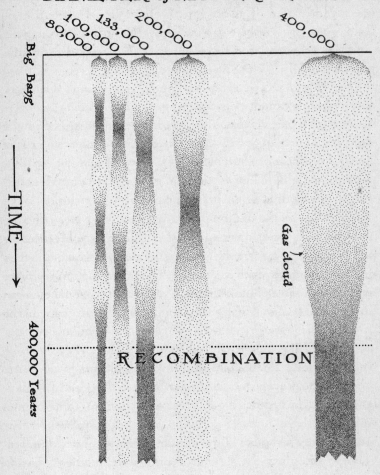

Acoustic oscillations: how gas expands and contracts before recombination

Recombination suddenly ended this tug of war, 400,000 years after the big bang. Electrons finally combined with nuclei, and the plasma became a transparent gas. Photons now passed through the gas unhindered and didn't scatter often — and they didn't transfer energy to the atoms. Thus, they could no longer give the atoms a kick through radiation pressure. Gravity finally got the upper hand, and the great clumps of matter fell together and formed galaxy clusters, galaxies, stars, and planets. Though the acoustic oscillations stopped at the moment of recombination, cosmologists realized in the 1970s that those oscillations in the early universe caused anisotropy in the cosmic background radiation. At the moment of recombination, clumps of matter were caught at different stages of collapse and expansion. Some clumps were fully collapsed, just about to expand again; they were hot and dense, and glowed brightly with radiation. Some clumps were fully expanded and just about to collapse again; they were cold and rarefied and glowed dimly. Most clumps were somewhere in between. But there was one additional element that made the acoustic oscillations — and the anisotropy in the cosmic background radiation — a very powerful tool for understanding the universe: the speed of light.

According to Einstein's theory of relativity, you cannot send any sort of information faster than the speed of light — you can't affect a distant body in any way until light has had the time to pass between you and that body — no matter how hard you try. We are eight light-minutes away from the sun, so if an evil genius such as Lex Luthor somehow made the sun blink suddenly out of existence, we earthbound humans wouldn't be affected until eight minutes later. For eight blissfully ignorant minutes, we would still see the sun shining in the sky, and the Earth would still feel the gravitational pull of the star and would continue its orbit, unaware that the sun was no more. According to Einstein's theory, the pull of gravity moves at the speed of light, and since light takes eight minutes to travel from the sun to Earth, so too the influence of gravity must take eight minutes to travel from the sun to Earth.

Applying this notion to the early universe has an interesting consequence. Since the universe was only about 400,000 years old when recombination happened and the acoustic oscillations ceased, any atom can feel the gravitational influence of matter only within a 400,000 light-year radius around it. Matter more distant than that (a few times the width of our galaxy) might as well have not existed as far as the oscillations are concerned; matter more than 400,000 light-years away could not have influenced the atom at all. (In physics-speak, the atom is not "causally connected" to anything more than 400,000 light-years away.) This means that a cloud of matter collapsing under its own gravity must have a maximum size of about 400,000 light-years across; if it is bigger, the more distant bits simply cannot feel the pull of some regions of the cloud. In other words, there is a maximum size to the collapsed clumps of matter; the hot spots in the cosmic background radiation cannot get beyond a certain characteristic diameter in the sky. Princeton's P. J. E. Peebles calculated that those hot spots should be about one degree in size, about twice as wide as the full moon in the sky.

These maximum-size clumps would have had just enough time to collapse before recombination hit. As a result, they glowed hotter than their surroundings, and astronomers should be able to spot these hot patches in the sky, fourteen billion years later. But not all clouds were as large as the biggest ones. Some clumps were a bit smaller, about half the size of the maximum clumps. These clumps would have collapsed in half the time, taking about 200,000 years to become maximally dense. But since recombination was still about 200,000 years away, the story does not end there. When one of these smaller clouds was collapsed as far as it could go, it glowed hot, and radiated an enormous amount of light. That intense light pushed the matter apart, forcing the cloud to expand again. As the cloud grew larger, pushed apart by radiation pressure, the matter would have extended to its maximum size just as recombination hit. Since these clouds were sparse, they would be cooler than surrounding matter—these half-size clouds would show up as cold spots in the cosmic background radiation. Yet smaller clumps, about a third of the size of the maximum clumps, would have had time to collapse, expand again, and collapse once more in the 400,000 years before recombination. These one-third-size clouds were completely collapsed—at their maximum density—at recombination, so they would be hot spots in the sky.

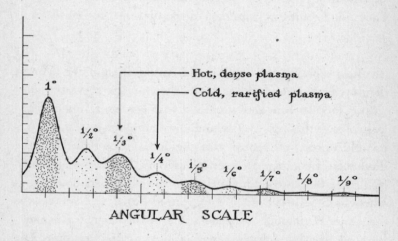

Peaks and valleys in the power spectrum of the cosmic background radiation

The cosmic background radiation, argued the theorists, should be speckled with hot and cold spots—the "fundamental" hot spot should be about one-degree in size, with hot spot "overtones," of one-third of a degree, one-fifth of a degree, and so forth. It should also have cold spot overtones at one-half a degree, one-quarter of a degree, and so on. To cosmologists, the hot spots are peaks in a graph, and the cold spots are valleys. Peebles calculated the positions of those peaks and valleys; the first peak, the fundamental, the clouds of maximum size, should be at one degree; the second peak should be at one-third degree, and so forth. Unfortunately, at the time nobody had any way of measuring such tiny features. The COBE satellite's vision could only see features larger than seven degrees across, and the biggest hot spots are one degree wide. Worse yet, astronomers did not know for sure that the peaks and valleys would be at their assigned places, because the very nature of space and time can distort astronomers' vision, making the features even harder to see.

Cosmologists had to wait another decade after COBE to see the first peak in the cosmic background radiation. When they finally did, they confirmed the ultimate fate of the universe.

In the perpetual daylight of the Antarctic summer, Mount Erebus isn't as forbidding as its demonic name would suggest. Clad in white and wreathed with clouds, it juts majestically above the frigid wasteland. To cosmologists, though, it is merely a prop in the most stunning photograph that they have ever seen, a composite image released by the Boomerang team in April 2000. Superimposed upon the sky, in shades of psychedelic blue, the cosmic background radiation blankets the heavens. Hot spots and cold spots in the microwave background, invisible for years, were visible for the first time. The fate of the universe was written in the sky, and cosmologists could finally read the writing.

The first instrument with vision fine enough to see the hot and cold spots in the cosmic background radiation was an ungainly metallic-looking "telescope" known as Boomerang (a contorted acronym of Balloon Observations of Millimetric Extragalactic Radiation and Geophysics). Boomerang dangled from an enormous helium balloon above the frigid Antarctic in the summer of 1998. Working in the Antarctic is expensive and cumbersome, not to mention cold as heck. However, two important reasons drew the Boomerang team, an international collaboration of thirty-six scientists, to the antipodes. The Antarctic is the coldest place on earth, and as a result it radiates the least energy—its ground radiates the fewest microwaves, and microwaves can swamp out the weak signal from the farthest reaches of the universe. The colder the instrument and the colder the surroundings, the better the chance of seeing the tiny fluctuations in a faint hiss of microwave noise. But there was another reason that made the Antarctic an ideal laboratory for cosmic microwave background measurements: a quirky wind current. There is a stream of wind that circles the Antarctic. If you stand in the right spot and release a balloon into the atmosphere, the balloon will get caught by the wind, circle the pole, and return to its launching spot a bit more than a week later. This is precisely what the Boomerang scientists did with their telescope.

Boomerang is an extremely sensitive machine designed to pick up microwave radiation from the sky without confusing it with energy from the ground or even the instrument itself. The machine gathers light from the sky and guides it onto small *bolometers,* heat sensors that can pick up even the weakest signal from the microwave sky. The bolometers were suspended in a spiderweb of filaments to isolate them from heat and to make them particularly sensitive to microwave light. They worked beautifully.

COBE's seven-degree resolution was too coarse to pick up the tiny hot and cold spots in the cosmic background radiation. Boomerang, on the other hand could resolve spots as small as one-third of a degree across. No longer were scientists looking at the cosmic background radiation with blurry glasses. They were beginning to read the writing on the walls. That writing told of the fate of the universe. It also told of its shape.

Yes, shape. Though it might seem as ridiculous to discuss the universe's shape as to try to figure out the universe's scent, it makes perfect sense to mathematicians and physicists. The equations of general relativity liken space and time to a flexible fabric, something like a rubber sheet. Mathematicians have a set of tools that they use to describe curvy and stretchy objects. They make up a whole field of study called differential geometry. Differential geometry allows mathematicians to study curves and surfaces in space. It allows mathematicians to analyze quantities like curvature and torsion that describe the properties of an object in space. Though a rubber-sheet spacetime seems like an artificial construct, it is a very natural and powerful idea when you have the tools to deal with it.

Einstein's key insight, the idea that formed the basis for the general theory of relativity, was that space and time behave, mathematically, like a smooth surface. This has a few important consequences. For one thing, it explains where gravity comes from. A heavy object, like our sun, distorts that spacetime fabric, bending it slightly, like a bowling ball on a mattress. If you place a marble on the mattress, it will roll toward the bowling ball because of the curvature of the mattress. Likewise, if you place an asteroid near the sun, it will fall toward the sun, because the curvature of spacetime forces it to move in that direction.

Why space and time, rather than just space? Einstein realized that your motion in the everyday three dimensions of space (up-down, left-right, and back-front) also affects your motion through the fourth dimension, time. For instance, if you move very, very fast in space, your wristwatch will tick very, very slowly with respect to your clock back on Earth. Though space and time have slightly different mathematical properties (our four-dimensional universe has three "spacelike" dimensions and one that is "timelike"), they are inseparable. Affect space and you automatically affect time, and vice versa. So in a mathematical sense they are woven together.

Since space and time are like a fabric, spacetime can have a curvature locally, like the distortions caused by the sun, or "globally," a curvature for the entire universe.[8] It is something like our own Earth. Locally, the surface of the Earth

---

8. This is a good example of how language fails when dealing with something as enormous as the universe. *Globally* used to be the biggest thing we could think of, which seems silly when we are talking about the cosmos as a whole.

has peaks and valleys, rolling hills and crevasses, little lumps and divots that affect a small area on the surface. But zoom out far enough and you see that the Earth is a sphere, even though the curvature is all but imperceptible across small distances. It is the same with the universe as a whole. Locally, the fabric of spacetime can be flat, or it can have ripples; it can even have immense, seemingly bottomless pits. The universe as a whole, however, also has a shape. It might be flat, or it might have *positive curvature* like a ball, or *negative curvature* like a saddle. All of these shapes are in four dimensions, of course, so they are very difficult to visualize, even with training. Nonetheless, the three-dimensional versions—a plane, a sphere, or an enormous saddle—are reasonable approximations of what is happening in our 4-D universe.

The shape of the universe, according to the equations of general relativity, is closely tied to the amount of "stuff"— matter and energy—that the universe contains. Einstein's equations state that matter curves the fabric of spacetime, and the more matter there is in the universe, the more highly curved the universe, as a whole, is. If there is more than a critical amount of matter, then the universe has positive curvature, like a ball. If there is less than this critical amount, then the universe has negative curvature, like a saddle. If the amount of stuff in the universe is precisely enough to balance it on the knife-edge between positive and negative curvature, the cosmos is flat, like a plane. Scientists use the symbol Ω (the Greek capital letter omega) to represent the amount of stuff in the universe. The size of Ω determines the curvature of the universe; if it is below the *critical density,* if omega is less than one ($\Omega < 1$), then the universe tends to have negative curvature and is shaped like a saddle. If omega is greater than one ($\Omega > 1$), then the curvature is usually positive, and the universe has a surface like a ball. If $\Omega = 1$, then the universe is pretty much flat, like a plane.

The symbol Ω is a particularly apt choice, because the curvature of the universe is related to its fate. Omega, the last

letter of the Greek alphabet, symbolizes the end of everything, just as alpha symbolizes the beginning. Omega is a
measure of the stuff in the universe, the matter and energy
that make up the cosmos, and omega determines the victor in
an eternal struggle: the struggle between expansion and con-
traction, the battle between an ever growing universe and one
that collapses under its own weight. For just as Ω determines
the curvature of the universe, it is related to how the universe
dies. If it is below the critical density, if $\Omega < 1$, then there is
not enough stuff to counteract the expansion of the universe,
and the cosmos expands forever and dies an icy death. If
$\Omega > 1$, then there is more than enough stuff to overcome the
force of the initial explosion, and the universe's expansion
stops, reverses itself, and leads to a fiery big crunch. The case
of $\Omega = 1$ is special: The universe dies a cold death, as the expansion never quite ceases.[9]

Curvature, the amount of stuff in the universe, and the
ultimate fate of the cosmos are all interrelated.[10] Determine
one of them and you can divine the other two. And the cosmic
background radiation gave cosmologists a way to measure
the curvature of the universe directly.

Einstein's theory of general relativity states that light
doesn't necessarily travel in straight lines; instead, it follows
contours of the surface of spacetime, called geodesics. On a
flat plane, geodesics happen to be lines. This means that two
ants, marching on parallel lines, will always stay the same distance apart. Likewise, in a flat universe, two parallel light

9. Cosmologists liked the idea of having a flat universe because of mathematical simplicity, but all their measurements—grist for a later chapter—implied that the total
amount of matter in the universe was roughly a third of what was needed to have a
flat universe.
10. Before scientists really considered the possibility of a cosmological constant, Λ,
the curvature of the universe determined its fate; a positive curvature meant that the
universe would collapse in a big crunch, while a negative or zero curvature would
mean that the universe would expand forever. But Λ messed up this nice, neat cor-
respondence, and it is possible, once you allow for Λ, to have a positively curved universe that expands forever or a negatively curved one that collapses. Nonetheless, the
concepts are still related, though their relationship is more complicated than was once
thought.

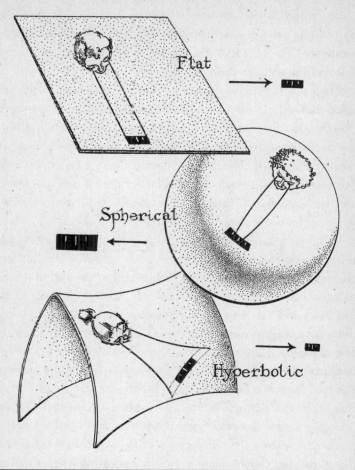

How the curvature of spacetime distorts the images of distant objects

rays will always stay the same distance apart as they approach an observer. But on a surface with positive curvature, the term *parallel* doesn't make any sense. Geodesics on a sphere are great circles, like lines of longitude. If two ants at the North Pole began marching down lines of longitude, they would start out inches away from each other and wind up separated by miles and miles. In a positively curved universe, this effect distorts the apparent size of distant objects; incoming rays, in a sense, are spread farther apart so objects look bigger than they normally would. On a surface with negative curvature, like a saddle, the opposite is true, making distant objects look smaller than usual.

This suggests a way to figure out the curvature of the universe. All you need is a standard ruler. Take an object of known size and transport it a great distance away — halfway across the universe. Compare its apparent size to the size you expect it to be; if it looks smaller than you expect, then the universe is saddle shaped. If it is larger than you expect, then the universe has a positive curvature; it is shaped like a sphere. The only trick is to find that standard ruler.

That is exactly what Boomerang did. The fundamental hot spots in the cosmic microwave background are standard rulers. Theorists knew precisely how big these hot spots are supposed to be, based upon how far light can travel in the 400,000 years between the big bang and recombination. These hot spots are effectively blotches of known size on the most distant object that astronomers will ever see. Since they have a known size, these hot spots are standard rulers. If the universe was flat, theorists expected those blotches to be about one degree wide. If the universe was curved like a sphere, then those splotches would appear bigger than expected, maybe a degree and a half or two degrees across. If the universe was shaped like a saddle, the hot spots would appear smaller than expected, perhaps two-thirds of a degree or half a degree wide.

The fundamental hot spots were precisely as big as theorists expected them to be: one degree wide. This means that the light from the distant universe is not distorted by the shape of spacetime; the universe is not curved like a sphere or curved like a saddle. Boomerang's data provided powerful evidence that the universe has no curvature. The world might be round, but the universe is flat.

This was powerful support for the supernova hunters' conclusions. Mathematically, a flat universe tends to expand forever; the supernova data also suggested that the universe will expand forever, as the expansion is getting faster and faster rather than slower and slower. The two results, each using different techniques, support each other, leaving little wiggle room for scientists who question the individual results. Cosmologists were forced to agree. For the first time in the history of humanity, we know how the universe will end. We have answered one of the questions that has plagued philosophers since the beginning of civilization. We are now almost certain that the universe will die a death by ice.

But just as the flat universe confirmed the fate of the cosmos, it also compounded the dilemma that the supernova hunters had caused. When the supernova hunters saw that the universe was expanding ever faster, they realized that there was no known way of accounting for that acceleration without assuming the existence of some repulsive force like the cosmological constant, Λ. It was a dramatic conclusion, and many scientists thought that something had to be wrong with the supernova data. However, Boomerang's measurement of the shape of the universe, and of Ω, showed that the supernova data were right. The measurement of Ω is a census, a measurement of the amount of stuff in the universe—all the matter and all the energy. Boomerang's data showed that there was something else in the universe besides matter; something else had to be flattening out the universe. The existence of Λ, the cosmological constant, or some other form of mysterious "dark energy" seemed to be written on the very walls of the universe. There was no escaping it. Scientists had to admit that the vast majority of stuff in the universe was unknown to science. A mere half-decade ago, nobody would have dreamed that cosmology would be in such a state. The third cosmological revolution was fully under way.

Chapter 6
The Dark Universe

[WHAT'S THE MATTER WITH MATTER?]

O first created Beam, and Thou great Word,

"Let there be light, and light was over all";

Why am I thus bereav'd thy prime decree?

—JOHN MILTON, *SAMSON AGONISTES*

Cosmology has entered an era of stunningly precise measurements. No longer do cosmologists have to be satisfied with coarse estimates of the fundamental properties of our universe; they can look up in the sky and measure them with caliper-like precision. But that precision is creating a troubling and surprising picture of the cosmos. "It's an absurd universe," says Michael Turner, a cosmologist at the University of Chicago.

The absurdity is encoded by a single symbol: Ω, the amount of stuff in the universe.[1] With their new, superprecise

1. In truth, Ω refers to the density rather than the amount of stuff, but we can use the two ideas interchangeably.

tools, cosmologists and astronomers have finally made the first measurements of Ω. Scientists are finding out in incredible detail what the universe is made of, how it was born, and how it will end. The answers are baffling and are forcing scientists to rethink their theories about the nature of the universe.

The supernova data and the cosmic microwave background data showed that $\Omega = 1$, that the universe is flat and that it will expand forever. While this was a major victory for cosmology, it posed an incredible problem, because at first glance there does not seem to be nearly enough stuff to make a flat universe.

There are two components to Ω: matter and energy. Assessing the matter part should be relatively straightforward; scientists have long been studying the properties of matter, and the stars and galaxies that dot the heavens are all made of matter. Nevertheless, when cosmologists tallied up all of the matter that they could see, everything in all the galaxies visible to even the most powerful telescopes, there was way, way too little stuff to make $\Omega = 1$. So when the cosmic background measurements showed beyond a reasonable doubt that $\Omega = 1$, cosmologists were forced to admit that they did not know what most of the universe was made of. It was rather embarrassing.

The mystery is twofold. The supernova data showed that an unknown force or *dark energy*—perhaps Einstein's cosmological constant—is causing the universe to expand ever faster. If that isn't bad enough, the new picture of the universe also implies that a baffling, unseen component of matter, *dark matter*, is largely responsible for the way the universe evolved. Nobody has seen dark matter; nobody has trapped dark matter in an earthbound laboratory. Nobody really knows, in detail, what the properties of dark matter are. Yet most cosmologists are absolutely convinced that both dark matter and the strange antigravity dark energy exist and shape the universe. More troubling still, they assert that dark matter vastly outweighs the ordinary matter in the universe, the stuff that makes up stars, planets, and people.

As farfetched as this cosmological picture seems, all astronomical measurements—of cosmic background radiation, of the distribution of galaxies, of distant supernovae, of the proportions of different types of matter in deep space—are forcing cosmologists to accept that two-thirds of the universe is invisible, and that most of the cosmos is composed of material that humans have never seen and never measured. The portrait of the universe that is coming into focus seems to have been painted by a surrealist. "We may be made of starstuff, but the universe ain't," says Turner.

At this point, you are probably extremely skeptical about the picture of the universe that I am describing. You should be. No good scientist accepts a statement merely on faith, and no reader should either. But step by step, cosmologists came to accept the new picture of the universe, and I will retrace their journey to convince you too.

Omega will be our guide. We will investigate each component of Ω, piece by piece. The first component is matter. Matter is the face of the dark universe.

The story of matter is nearly as old as the universe itself. In the period just before nucleosynthesis, a period that began about a millionth of a second after the big bang and lasted for a few seconds, the universe was filled with protons and neutrons that were moving so fast that they could not stick together. These particles, relatively heavy for subatomic particles, are known as *baryons*.[2] Almost all the matter we encounter in everyday life, the stuff that makes up objects on Earth, is mostly baryonic, because it is pretty much made up of protons and neutrons. (The light electrons contribute very little to the overall mass of stuff on Earth.) A few seconds after the big

2. In Greek, the word *barys* means "heavy." Leptons are lightweight particles like electrons; the Greek *leptos* means "small." Middleweights, like the pion, are known as mesons. You've probably already guessed that *mesos* means "middle."

bang, the universe cooled enough so that the protons and neutrons could stick to one another. Nucleosynthesis began; some of the protons and neutrons collided and stuck together, forming elements heavier than hydrogen, such as helium.

The nucleus of every atom is made up of protons and neutrons, though the nucleus of the simplest atom, hydrogen, is a proton all by itself. The proton is extremely stable, so it can sit by itself for all eternity without breaking apart.[3] The neutron, on the other hand, is not nearly as stable. Left to its own devices, it will decay in about fifteen minutes, turning itself into a (slightly lighter) proton and spitting out an electron. If neutrons created shortly after the big bang had not occasionally slammed into a wayward proton or two, there would be no neutrons at all; all the baryons in the universe would be protons. Consequently, all the baryonic matter created in the moments after the big bang—the "primordial" baryonic matter—would be hydrogen. But this is not the case. About 25 percent of the primordial baryonic matter in the universe seems to be helium.

Nucleosynthesis saved neutrons from extinction. When a neutron smacked into a proton, the two stuck together, forming a heavier nucleus, deuterium. Though deuterium is rather fragile, unlike a lone neutron it does not spontaneously decay. Because deuterium is stable, some of the deuterium formed in the era of nucleosynthesis is still around today. Thus, there was deuterium in the primordial baryonic matter as well as hydrogen, but the story of nucleosynthesis does not end there. During the melee of nuclei in the first few minutes of creation, a proton sometimes smacked into a deuterium nucleus, forming helium-3. Hit by another neutron, helium-3 becomes helium-4. All of these elements, and more, were created during the era of nucleosynthesis, the time when the universe was hot and dense enough to sustain nuclear fusion. But a few minutes after the big bang, the universe expanded and

3. As far as we know. There is a possibility that it decays over a very, very long time.

cooled and the furnace shut down. Protons, neutrons, and the newborn atomic nuclei didn't have enough energy to stick together when they slammed into one another. No longer could cruising protons or other nuclei pick up additional neutrons or protons; the era of big bang nucleosynthesis was over. The primordial baryonic matter in the universe was fixed forever. Most of it, about 75 percent, was hydrogen, and almost all the rest was helium, with a few other trace elements.

In the 1940s, George Gamow figured out that the proportion of hydrogen to helium, deuterium, and the other elements in the primordial matter of the universe was closely related to the density of matter in the early universe.[4] Imagine, on one hand, that baryons were relatively rare in the early universe, that there was a relatively large amount of space between baryons during the era of big bang nucleosynthesis. If this were the case, then protons and neutrons would be very unlikely to slam into each other, because there is so much empty space in between. It is something like wandering in the Alaskan wilderness; it is unlikely that you will randomly bump into another person—there just aren't all that many people around for you to smack into. So, if the density of baryons in the early universe was small, protons and neutrons would not bump into each other very often, and they would produce very little deuterium and helium. An astronomer looking at primordial gas clouds, the almost pristine stuff that floats in nearly empty intergalactic space, would notice that the clouds were made almost entirely of hydrogen; they would have very little helium and deuterium, because very little of the stuff got made before the era of nucleosynthesis ended.

On the other extreme, if the density of baryons in the early universe were extremely high, if the protons and neutrons were crowded together like commuters in a busy Tokyo subway, they would be slamming into each other all the time. All the collisions would produce a lot of deuterium, helium,

4. The calculations establishing this were the ones that drove his colleagues Ralph Alpher and Robert Herman to the first prediction of the cosmic background radiation.

lithium, and other even heavier elements, before the temperature dropped too far to allow any further agglomeration. A lot of the hydrogen would have been converted into heavier elements. This time, in a high-baryonic-density universe, an astronomer would notice much more helium and other elements in the primordial gas clouds and less hydrogen.

Gamow realized that this logic can allow a clever astronomer to figure out the conditions in the early universe by measuring the relative amounts of hydrogen to helium and other elements in primordial gas clouds. A large proportion of hydrogen and a small proportion of helium would mean that the density of baryons in the early universe was small. A smaller proportion of hydrogen to a larger proportion of helium would mean that the baryonic density was relatively high. Gamow understood that by looking at the ratio of hydrogen to helium and other elements in those primordial gas clouds, cosmologists could, in theory, calculate the so-called baryonic fraction in the early universe.

Recall that the symbol Ω describes the density of stuff in the universe. Einstein's equations relate the density of stuff to the curvature of the universe, so, by convention, cosmologists defined $\Omega = 1$ to mean that the density of stuff in the universe is just enough to make spacetime flat. If $\Omega > 1$, the universe has positive curvature, like a sphere, and if $\Omega < 1$, the universe has negative curvature, like a saddle. Though it might be confusing at first sight, cosmologists use the symbol Ω_b to represent the baryonic component of Ω. That is, if all the stuff in the universe were baryons, Ω_b would equal Ω. But as it happens, there's more to the universe than just baryons, so Ω_b is just one component of Ω.[5] In any case, cosmologists can figure out what Ω_b is by looking at the proportion of hydrogen to helium-4 and other elements in primordial gas clouds.

5. The other components are discussed in future chapters. Also, cosmologists compensate for the expansion of the universe by multiplying Ω and related quantities by a factor that takes the Hubble constant into account. For simplicity's sake, I have consistently omitted that factor.

Lots of hydrogen and little helium-4 means a small Ω_b, and the less hydrogen and more helium-4 there is, the larger Ω_b must be.[6]

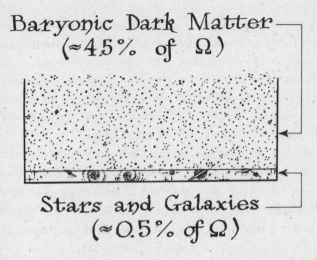

Baryonic Dark Matter
($\approx 4.5\%$ of Ω)

Stars and Galaxies
($\approx 0.5\%$ of Ω)

Using sensitive spectrometers, astronomers analyze the light that passes through clouds of primordial gas, and by looking at the fingerprint of the light that the gas absorbs, they can figure out the proportions of the various elements in the clouds. Over the years, their measurements have gotten better and better, and they have a pretty good idea of what the ratios of hydrogen to heavier elements are in those gas clouds. From those measurements, they've calculated an Ω_b of 0.05 or so. That is, baryonic matter, taken by itself, makes up about 5 percent of what is needed to keep the universe flat.

There is another way of totaling up the baryonic matter

6. For elements other than helium-4, the story is more complicated. For instance, the proportion of deuterium goes *down* with increasing density, because deuterium is consumed when helium-4 is created; when less helium-4 is created, more deuterium is left intact. Observations show that the abundances of all these elements are incredibly consistent with the predictions, and with the theory pioneered by Gamow, providing yet another very strong confirmation of the big bang theory independent of the cosmic background radiation. Big bang theory says that, under any given condition, you have to have certain proportions of hydrogen to helium-3 to helium-4 to deuterium to lithium, and that is precisely what astronomers see. Yet another big bang prediction turns out to be true. Nucleosynthesis, the Hubble expansion, and the cosmic microwave background are pillars that support the big bang theory.

in the universe. Baryonic matter is the ordinary matter that we're all used to; the stuff of atoms, stars, and galaxies, the stuff that shines brightly, is all baryonic matter. Thus, by estimating how many galaxies are in the universe and estimating how much matter each of those galaxies contains, astronomers hoped to get another total. But this method came up short. Very short. If you add up all the visible matter in the universe, you get a value of about 0.005—about one-tenth of what you would expect based upon big bang nucleosynthesis. How to account for the discrepancy? Dark matter.

The idea that most matter in the universe is dark—that is, invisible to telescopes—is naturally a disturbing one. How can you spot an invisible object? Yet scientists have seen it. To find dark matter, instead of looking for it directly, they look for what it does. They look for its gravitational attraction, the mutual pull between two objects that have mass.

Physicists have understood gravity pretty well since the seventeenth century. In 1687, Isaac Newton's magnum opus, the *Principia*, described the law of universal gravitation, which has survived for half a millennium with very little modification. At the beginning of the twentieth century, Einstein's theory of relativity extended Newton's theory; however, except for objects that are in very strong gravitational fields or that are moving very fast, Newton's laws do a nearly flawless job describing the motion of objects under the influence of a gravitational field.

They work beautifully for our solar system. If you know the mass and position of the objects in the solar system, you can calculate the direction and the strength of the pull of gravity on any given object, and how it will move.[7] The Earth

7. The only exception is the planet Mercury, which deviates slightly from the Newtonian prediction. Thanks to Einstein, we now know that the mass of the sun warps the space and time around it. Mercury, which is very close to the sun, feels the curvature of spacetime in a way that Newton never predicted, altering the orbit of Mercury slightly.

is about 93 million miles from the sun, so that tells us how strongly the sun pulls on the Earth, and how fast it must move in its orbit around the sun: about 18.5 miles a second, perfectly in accord with observation. Jupiter, at a much more distant 483 million miles from the sun, feels less pull and orbits the sun at only about 8.1 miles a second. Neptune, at more than 3 billion miles from the center of the solar system, feels less pull still and pokes along at a mere 3.4 miles per second, taking nearly 165 years to complete one orbit around the sun. The farther away an object is from the sun, the less pull it feels. The less pull it feels, the slower it moves around the center of the solar system. It's a fact as strong as the Newtonian laws of gravity and motion — it's just the way solar systems (not just our own) behave.

Not only do these laws apply to solar systems, they apply to any disk-shaped object, like a spiral galaxy. The farther away from the center of a galaxy a star is, the less it feels the gravitational tug of the other stars pulling it toward the galactic center. The farther a star is from the center of a disk-shaped galaxy, the slower it should orbit the galactic core. This too should be as solid as Newton's laws. Alas, it simply isn't true. In the late 1960s, Vera Rubin, an astronomer at the Carnegie Institute of Washington, pointed two telescopes at the nearby Andromeda galaxy in an attempt to measure how fast the stars were moving around the center of the galaxy. Like Hubble atop Mount Wilson, Rubin and a colleague measured the fingerprint-like spectrum of lines given off by elements like hydrogen, and whether the spectra were Doppler shifted toward the red or the blue end of the spectrum. This revealed how fast the stars were moving toward or away from the Earth, and how fast the stars were spinning around the center of Andromeda.

Newton's laws would seem to imply that stars near the core of the Andromeda galaxy should move relatively quickly; the farther a star was from the nucleus of the galaxy, the slower it should move. Rubin's observations told a different story. In 1970, she showed that however far the stars were from the center of the galaxy, they moved at about the same speed, approximately 150 miles a second around the center of Andromeda.[8] Newton's predictions, amazingly, were not holding up. Something seemed to be wrong with one of the fundamental laws of nature. Rubin's observation was not a fluke, either; when astronomers looked at other galaxies, they noticed the same pattern. Stars all spin at roughly the same speed around the center of the galaxy, defying standard Newtonian motion, which states that stars must move slower and slower toward the edges of the galaxy.

Newton's laws were absolutely wrong in galaxy after galaxy. This meant that either Newton's laws had to be discarded, or disklike galaxies weren't disklike after all. Few astronomers wanted to toss out Newton, but how could a disk-shaped galaxy not have a disk shape? When astronomers look at a spiral galaxy, they see a disk. If they count the stars, one by one, estimate their masses, and use Newton's laws to calculate their motions, the galaxy should behave like a disk. If it looks like a disk, and Newton says it should behave like a disk, how could it not be a disk? There is a way: if there is more to the galaxy than the visible stars. Galaxies must contain invisible dark matter.

If there is additional matter in the galaxy that is dark rather than shining brightly like the stars that we observe, we would not see it—but it could explain the strange rotational behavior of the stars. A cloud of dark matter surrounding the galaxy, a dark halo invisible to telescopes, would still have mass and gravitational attraction. This dark matter halo would provide a little extra pull on far-out stars, making them move faster than we would expect, since our calculations only took the shining stars that we see into account. Indeed, if galaxies were surrounded with a halo of dark matter, the outermost stars would

8. The title to Rubin's paper referred to Andromeda as a nebula rather than a galaxy, hearkening back to the pre-Hubble name. Old nomenclatures die hard.

move just about as fast as the innermost ones, which is precisely what Rubin and other astronomers saw. Though Rubin herself believed that dark matter halos were responsible for the non-Newtonian motions of stars in galaxies, most astronomers found the idea difficult to swallow. Some physicists, in fact, preferred to reject Newton's laws and abandon the well-worn relationship between the force of gravity and distance rather than accept an enormous shroud of unseen matter about every galaxy in the universe.

Rejecting Newton is a radical solution to the galactic-motion problem, but it is not that much worse than proposing a shroud of dark matter around every galaxy. By modifying the equations of Newtonian gravity slightly, making the force of gravity at great distances a little stronger than Newton's laws suggest, you can explain the extra speed of galactic stars without a cloud of dark matter. The best of these theories, according to gravitational physicists, is called MOND, shorthand for Modified Newtonian Dynamics. Invented in 1983 by Mordechai Milgrom, a physicist at the Weizmann Institute of Science in Israel, MOND's equations give a stronger pull to gravity at large distances than does Newton's law of gravitation. That extra strength gives stars a little additional pull if they are far from the center of a galaxy, which, in turn, causes them to wheel about the galactic nucleus with a little extra speed. MOND does an excellent job at explaining the velocities in galaxies, but it does a poor job of predicting the speeds of galaxies that gyrate around the center of a galaxy cluster, so it had some serious problems from the very start. The universe is littered with massive galaxy clusters, each containing hundreds of galaxies spinning around one another.

Our own galaxy, as well as Andromeda and other nearby galaxies, is part of the Virgo supercluster, so named because the center of the group is located in the direction of the constellation Virgo. Galaxies orbit the center of a cluster at much, much greater distances than stars orbit the nucleus of an individual galaxy. This means that the gravity-enhancing equations of MOND would give an incredible kick to those orbiting galaxies, and to the gas trapped within a cluster; they would be spinning extremely fast around the center of the galaxy cluster. But recent observations of galaxy clusters, and of the gas within them, do not support MOND's predictions. Even Milgrom himself admits that MOND has serious problems with explaining the motions in galaxy clusters. "There is a conundrum," he says, and suggests that additional unseen matter might account for the discrepancy. "There is always room for yet-undetected matter," he says. That means that MOND needs dark matter too—and the whole reason for MOND was to eliminate the need for dark matter. Though Milgrom still holds fast to MOND, he admits to the possibility that his theory will one day be falsified. "As its inventor, I would like it to be a revolution, but I look at it coolly," he says. "I will be very sad, but not shocked, if [the answer] turns out to be dark matter."

Since even the best alternative to dark matter is in trouble—and, in fact, requires dark matter to make it work—cosmologists and astronomers are forced to accept the existence of some form of invisible matter to help hold galaxies and galaxy clusters together. As distasteful as it might be to believe in something that is invisible and, so far, undetectable, it is the best alternative.

If you accept the existence of dark matter, then it makes perfect sense that astronomers who counted galaxies came up short when trying to estimate the mass of the universe. After all, when you count all the visible galaxies, by definition you are only counting what you see. If there is a significant proportion of dark matter in the universe, as the galaxy-rotation curves lead scientists to believe, then the galaxy-counting method will vastly underestimate the amount of baryonic matter in the universe. This is precisely what happened, and it explains why Ω_b can be so much greater than the visible mass in the universe. The best evidence so far is that only about one-tenth of the baryonic matter in the universe shines with a light that we can see, and nine-tenths is dark matter. It all fits together.

Astrophysicists are forced to accept the existence of dark matter. They can only explain their observations of the universe if they invoke its presence. The motion of stars around the galactic nucleus, the orbits of galaxies around the center of a galaxy cluster, ratios of elements in primordial gas clouds — all these observations lead cosmologists to the inevitable conclusion that *most* of the matter out there is invisible. All attempts to explain the nature of the universe without resorting to dark matter have failed miserably. Perhaps, like almost all cosmologists, you're convinced; if not, there is more evidence to come. But be warned, the story is about to get even stranger. There is more than one type of dark matter. There is the ordinary stuff, the baryonic dark matter that is like the matter we encounter in everyday life. But there is also an exotic form of dark matter that is not made of baryonic matter at all. Physicists don't really understand what this exotic form of dark matter is, but they all agree that it exists. It is implausible and unsettling, and it is at the heart of one of the most pressing problems in cosmology.

Chapter 7
Darker Still

[THE ENIGMA OF EXOTIC DARK MATTER]

A cosmic philosophy is not constructed to fit a man; a cosmic philosophy is constructed to fit a cosmos.
— G. K. CHESTERTON, "THE BOOK OF JOB"

Naturally, astronomers assumed that baryons made up, by far, most of the mass in the universe. Of course, they knew that other types of matter were out there, such as the electron, which is a lepton. But because baryons are so much heavier than leptons or other commonly seen types of matter (the neutron weighs about two thousand times more than the electron), physicists thought that the nonbaryonic component was insignificant—"ordinary" matter was almost entirely baryonic. Boy, were they wrong.

As scientists delved deeper into the components of Ω, they realized that the everyday baryonic matter was a small fraction of the total amount of matter in the universe. In the notation of cosmologists, the amount of matter in the universe, Ω_m, is significantly larger than the amount of ordinary, baryonic matter in the universe, Ω_b. This is an enormous problem. Not only is most of the matter in the universe dark, it is also unlike any of the baryonic matter that we're familiar with. Most of the universe is made up of exotic stuff, unknown to science. This is even more difficult to swallow than the existence of dark matter.

There are very good reasons to accept the idea of exotic matter, just as there are for believing in dark matter. The idea of dark matter made scientists uncomfortable—the idea of an invisible component to the universe seemed unnatural—but they were forced to accept it. Galactic motions and the abundances of helium and other elements in primordial gas clouds tell scientists that a lot more matter is out there than they can see. There is yet another good reason for believing in the existence of dark matter: we exist. The young universe was not smooth and uniform. Matter and energy were not evenly distributed over the surface of spacetime. There were clumps and voids. In the cosmic microwave background, physicists see the early clumps as hot spots in the radiation, and voids as cold spots. But in the present-day universe, those clumps and voids have an entirely different appearance. Clumps, where matter collected and collapsed under its own gravity, became enormous clusters of galaxies. The voids, on the other hand, became vast, empty regions of space, dotted only by the occasional galaxy that formed out of the meager supply of gas in the region. This nonuniformity was necessary for the creation of our sun and Earth; if the early universe had been uniform, then our own Milky Way probably would not have formed. And these clumps and voids reveal the existence not only of baryonic dark matter, but of exotic dark matter as well.

It is hard to visualize the structure of the universe. On a scale of hundreds of millions of light-years—just a fraction of the size of the universe—stars and even individual galaxies shrink into insignificance. When astronomers map the universe on large scales, they represent each galaxy, each enor-

mous cluster of hundreds of thousands of stars, with a tiny
dot. To make matters more difficult, it is not easy to put each
dot in its correct place. Finding the distance to any given
galaxy is fraught with uncertainty. The Hubble relationship
between velocity and distance gives astronomers a pretty
good estimate of the distance to any object, but the devil is in
the details. Individual motions of galaxies and reddening of
light due to dust clouds can mess up the calculations. Not
until the late 1980s did astronomers begin to get a coherent
picture of the large-scale structure of the universe. The picture they were getting was not quite what they expected. Instead of galaxies peppered more or less evenly throughout the
universe, scientists saw vast voids where almost no galaxies
appeared and thin tendrils of galaxies that connected supercluster with supercluster. We are living in a Swiss cheese
universe.

A delicate balance of forces led to those voids and tendrils. Gravity is the force that dominates the universe over
long distances. It is always an attractive force; two objects, if
they have any mass at all, pull each other together. Depending on how much energy and how much attraction the two
objects have, they can be bound together, like the moon and
the Earth, or the Earth and the sun, or the sun and the stars
in the Milky Way. Or, if the gravitational attraction between
the objects is not enough to overcome their relative motions,
they can zoom away from each other, with their mutual influence diminishing to nothing over time.

The more influential gravity is, the clumpier the universe
should be. Imagine a cosmos where the force of gravity grows
much, much stronger than it is in our universe. The planets
would probably not be able to maintain their stable orbits
around the sun; they would be drawn in, and at least the innermost planets would be slurped up by the hungry star,
which grows fat as it swallows the extra material. There
would almost certainly be fewer planets (if any) circling a
more massive sun. In a similar way, galaxies would draw

countless suns ever closer to the gaping maw of the super-massive black holes at their cores. The most central stars would be swallowed up as the force of gravity increases; the galaxies would shrink, getting ever denser. Some of the binary stars that circle each other would smash together, forming larger stars.[1] There would be fewer stars, and galaxies would be dense and small, and filled with massive stars. Likewise, galaxies orbiting the center of a galactic cluster would be drawn together. Clusters would shrink, galaxies would collide, and the dense clusters would have few massive galaxies instead of a large number of smaller galaxies. There would be fewer clusters too, as the vast conglomerations of galaxies would merge and collapse. The matter in the universe would be concentrated in a few massive clusters, rather than lots of smaller ones. The mass in the universe, instead of being spread out, would be concentrated into a few large lumps. The universe would be very clumpy.

On the other hand, if you make gravity much weaker, the universe would look a lot less clumpy than it does today. Because matter would not attract other matter very strongly, clouds of gas would only rarely collapse into stars; stars would seldom collect into galaxies, and galaxies would not bind themselves into clusters. The objects that form would tend not to be gravitationally bound to one another. Instead of collecting into clumps, the matter in the universe would stay as a fairly uniform haze throughout the universe. When gravity's influence is small, the universe is a lot smoother.

In reality, the force of gravity isn't tunable—it's a fixed constant. However, gravity's influence upon the universe is not fixed; it depends upon how much matter there is. If there is lots and lots of mass in the universe, then the influence of gravity is very strong; even though the force of gravity is not any stronger than before, there is more stuff that is exerting

---

1. Of course, stars would tend to go supernova under the extra influence of gravity, but that is beside the point in this example.

that force, making it a more important component of the forces that shape the universe. Likewise, if there is less matter in the cosmos, then gravity becomes less important. Because a smaller amount of matter is exerting a pull on other bits of matter, very few objects become gravitationally bound to one another. Therefore, the clumpiness of the universe gives astronomers a handle on the amount of matter in the universe. They can measure Ω_m.

From the abundances of helium and other elements in the universe, scientists know that Ω_b is about 0.05. But it turns out that an Ω_m of 0.05 is way too small to account for the present clumpiness of the universe. In fact, to explain why galaxy clusters appear the way they do, scientists need an Ω_m of about 0.35—about seven times as large as the accepted value of Ω_b. Because Ω_b is much smaller than Ω_m, the baryonic matter in the universe is in the minority; much more nonbaryonic matter is out there than baryonic matter. Just as the matter astronomers see is vastly outweighed by the dark baryonic matter in the universe, the dark baryonic matter in the universe is vastly outweighed by dark nonbaryonic matter. And since "ordinary" matter, the stuff we are familiar with, like stars and people, is baryonic, most of the matter in the universe must be something else entirely, something exotic that scientists do not really understand. Cosmologists have proved that they know almost nothing about the dark universe, the vast majority of matter in the cosmos. It's another embarrassing revelation.

As strange as the emerging picture of a dark universe might be, new measurements of the distribution of galaxies in the universe support it. Two large collaborations are busy mapping out the location of hundreds of thousands of galaxies and other large objects in an attempt to pin down that distribution. It is not a very sexy task, but it is incredibly important, and scientists across the globe are willing to take their careers off the fast track to complete these two enormous sky surveys.

The first is known as the Sloan Digital Sky Survey; it

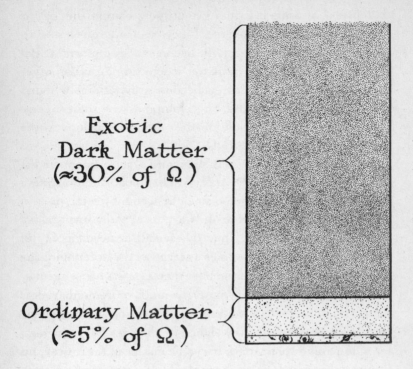

Exotic
Dark Matter
(≈30% of Ω)

Ordinary Matter
(≈5% of Ω)

uses a 2.5-meter telescope in the New Mexico mountains to map the skies. By the time the survey is complete in 2005, its scientists hope to catalog a million galaxies, revealing their position and distance.[2] However, to get the distance to a far-away galaxy, they have to record its redshift—its color—and if they expect to catalog that many objects in the sky, the Sloan survey cannot do them one at a time. So astronomers at the telescope measure the colors of each individual galaxy by drilling hundreds of holes in a metal plate. Each hole allows through only the light from one galaxy. The Sloan scientists

---

2. The calculations aren't always straightforward. The astronomers have to take all sorts of effects into account, such as the reddening of light due to dust, and the so-called finger of God effect, a distortion that stretches a distant galaxy cluster so that it looks like a long thin finger pointing directly at us. (There is no word on *which* of God's fingers it is.)

then use fiber-optic cables to pipe the light from the galaxy into a spectrum analyzer, which then reveals the distance to the galaxy. Drill another plate, make another observation. Drill another plate, make another observation. It is painstaking and difficult work. The scientists working on the rival Two Degree Field (2dF) project have it a little easier. They use a larger telescope, a four-meter instrument in Coonabarabran, Australia, and instead of drilling holes in plates, a robot places up to two hundred fiber-optic cables directly on the telescope. However, the 2dF survey will only map about 250,000 galaxies, rather than the million of the Sloan project.

After years of work, the first data from the teams filtered out. In April 2001 the 2dF survey released the data from the first 125,000 galaxies that it had mapped, and a few months later the Sloan Digital Sky Survey released its first results. Both sets of data imply that Ω_m is about 0.35. But they reveal even more than that. The same sets of data give an independent measurement of Ω_b as well.

The two surveys are providing exquisite pictures of the clumps and voids in the universe, the thin tendrils of matter that are surrounded on all sides by vast emptiness. These clumps and voids hearken back to the first 400,000 years of the universe, before the era of recombination. They are remnants of the time when pressure waves rattled through the plasma, compressing matter, letting it expand, and compressing it again. As the universe grew and cooled, the compressed regions gave rise to massive clusters of galaxies, while the rarefied ones stayed relatively free of matter. So by looking at the distribution of galaxies and voids, astronomers can figure out what the acoustic waves in the early universe were like, just as they can with the distribution of peaks and dips in the strength of the cosmic background radiation.

Different types of matter, however, transmitted those pressure waves in different ways. Ordinary baryonic matter would have compressed and expanded very strongly, as it was constantly being slammed about by the radiation pressure. "Exotic" dark matter, the kind believed to consist of nonbaryonic particles, barely interacts with light, so it does not feel the pressure of radiation as acutely as baryonic matter does. (That's what makes exotic dark matter invisible to telescopes.) Therefore, while the baryonic matter in the universe was getting buffeted by the light, the exotic dark matter would have oscillated more weakly, since it is almost immune to radiation pressure. Thus, wiggles in the distribution of galaxies—bumpy features in the graph that describe clumps of matter and voids on various scales—can reveal how strongly the matter in the early universe was oscillating, which, in turn, shows how much of the primordial matter is baryonic.

The 2dF and Sloan Digital Sky Survey both plot the distribution of galaxies in the universe, which, in turn, reveals the clumpiness of matter in the cosmos. Like the cosmic background radiation measurements, the data are presented in a hilly, bumpy plot quite similar to the plot of the cosmic background radiation spectrum (see page 81). Each peak represents a characteristic feature size. The 2dF team has claimed to see the wiggles in their plot. "That would be extremely exciting if they've seen it," says the University of Pennsylvania's Max Tegmark, though by the end of 2002 it was still too early to say for sure. Once they see those wiggles in the distribution of matter in the universe, they will be able to extract Ω_b. (Their preliminary results are very similar to the numbers that scientists get from nucleosynthesis.) The clumpiness of the universe is revealing how much dark matter is in the universe.

Looking at galaxy distributions, however, is not the only way to measure the clumpiness of the universe; scientists can just measure the clumpiness in the cosmic microwave background. Hot spots became clusters of galaxies, and cold spots became vast voids. So the peaks and valleys of the cosmic background radiation also give an idea of the clumpiness of the early universe—and how strongly the matter in the uni-

verse was being sloshed about by the intense light that buffeted the plasma. This, in turn, reveals the proportion of baryonic matter to other, exotic, matter. Everyone hoped that
Boomerang would allow cosmologists to get an ultraprecise
tally of the amount of stuff in the universe.

However, calculating the amount of matter in the universe required Boomerang's scientists to see more than just
the first peak in the cosmic background radiation. They
needed a second peak in the graph to figure out the ratio of
baryonic matter to exotic matter in the universe; they needed
to hear an overtone in the celestial music of the acoustic oscillations. When the first Boomerang data were revealed in
April 2000, the first peak was loud and clear, but the second
peak was conspicuously absent. It was as if the cosmologists
were expecting to hear the ringing of a bell, but heard the
bleat of a horn instead.

"The mischievous side of me wanted that to happen," said
Tegmark. For a time, physicists were scrambling to explain
what went wrong. A missing peak would mean that simple
models of how the universe formed and what holds it together
could not be correct. "You'd have to be violent to one of the
sacred cows of cosmology," Tegmark said. Everyone awaited
further data from Boomerang and other rival experiments; either the second peak would be found, or there was a major
problem with cosmological theories.

Luckily, the dilemma did not last too long. The first hints
of a second peak came from a telescope in Chile, known as
the Cosmic Background Imager (CBI). Unlike the Boomerang experiment, which used a balloon laden with sensitive
bolometers to detect the heat produced when microwave
photons strike them, CBI is a ground-based telescope that
uses *interferometry* to detect the microwaves from the early
universe.

If you look at a photograph of one of the most modern
military ships, like the missile cruiser USS *Ticonderoga*, the
first thing you will notice is a big, battleship-gray, monolithic

block that dominates the superstructure of the ship. It is ugly as heck, but it is the heart of the cruiser: an ultrapowerful radar that allows the vessel to track enemy ships, aircraft, and missiles for miles around. Unlike the radars of old, or the ones you see at an airport, it doesn't have to spin around to get a picture of the whole horizon. Thanks to interferometry, the cruiser's radar can point in any direction without moving a single mechanical part.

The key to interferometry is light's propensity to act like a wave.[3] Like waves in the ocean, a beam of light has crests and troughs. If two beams are lined up just right so that the crests of one are timed with the troughs of the other, and vice versa, the two will cancel each other—they interfere. On the other hand, they can reinforce each other if the troughs are lined up with the troughs and the crests with the crests. Light will appear bright where the waves reinforce each other and will disappear where they interfere with each other. You can see this for yourself if you hold two fingers together and look at a lightbulb through the tiny slit between the fingers. The tiny stripes you see (if you hold your fingers just right) are the places where the light waves, forced through the tiny slit, cancel out each other.

Instead of sweeping the beam of a radar dish across the sky, the USS *Ticonderoga* uses interferometry to point its beam. In place of a single large antenna that spins around, the *Ticonderoga* has a whole bunch of tiny antennas arranged in an array. Each antenna shoots out a beam of radar light, and by choosing the timing of the crests and troughs—the phase—of the little beams, the crew can control where in space the beams interfere and where the beams reinforce one another. By changing the phases of the individual elements, the crew can steer the area of reinforcement, the beam of the radar, about the sky without moving any of the antennas. Conversely, by adjusting the timing data of the little elements,

---

3. It can also act like a particle. In fact, it is both wave *and* particle.

they can "point" the receiver to any region of the sky they want, and ignore signals from regions of the sky from which they don't need information. CBI uses the same principle to concentrate on areas of the sky that are free from distractions; unlike bolometers, which do not have a good ability to discriminate where the light that strikes them is coming from, an interferometer can collect data from a very narrow patch of the sky, providing incredible resolution.

In some ways, working high up in the Andes is as difficult as working in the Antarctic. The workers have to make sure that they don't suffer from oxygen deprivation, and the station has to be entirely self-sufficient; oxygen, fuel, water, and all supplies must be brought up the mountain. Nevertheless, all the effort paid off. When CBI first pointed its interferometer at the heavens, it picked up hints of the missing second peak. Though nobody could say for sure that the second peak was there, it relieved some of the worries that Boomerang had sowed less than a year before. It also showed that the bumps and valleys in the cosmic microwave background tended to get weaker as the size got smaller, just as a musical instrument's overtones tend to get weaker as their frequencies get larger and larger. "The acoustic oscillations in the early universe were dying away. It shows we're on the right track, that the acoustic model is right," said Jeffrey Peterson, a cosmologist at Carnegie Mellon University in Pittsburgh. Better news was just around the corner.

A year after the Boomerang scientists released their first round of data, they were ready to reveal a much larger set of data. But their rivals were ready too. MAXIMA was a balloon experiment quite similar to Boomerang—the telescopes were almost identical—but MAXIMA had the disadvantage of flying over the continental United States rather than over the Antarctic chill. As a result, their data were messier. However, the Degree Angular Scale Interferometer (DASI), an instrument that worked in a manner similar to CBI, had been taking data in the Antarctic, and the DASI team members had finally crunched their first round of measurements. In April 2001, at a meeting of the American Physical Society in Washington, D.C., all three teams presented their latest results, and the missing peak was there in all its glory. "This is like Santa Claus is arriving," said Tegmark. Boomerang and DASI saw the first peak, the second peak, and the third peak. In June 2002, CBI released its second set of results, which contained the third, fourth, and hints of the fifth and sixth peaks. This allowed the teams to calculate Ω_b and Ω_m. The numbers they got agreed with all the other measurements: the baryons make up about 5 percent of the stuff in the universe, and all matter, taken together, makes up about 35 percent. (That is, $\Omega_b = 0.05$ and $\Omega_m = 0.35$.) These are the same values that scientists got when looking at the proportions of elements in the universe and the distribution of the galaxies.

All the measurements have come to the same conclusions. Big bang nucleosynthesis, galaxy maps, and the cosmic microwave background measurements show that the amount of ordinary, baryonic matter in the universe is only 5 percent of what is needed in a universe where $\Omega = 1$, yet the total amount of matter makes up about 35 percent of that stuff. The difference—about 30 percent of the stuff in the universe—should be a nonbaryonic, exotic form of matter. Scientists have not yet discovered a candidate.

Cosmologists are shaking their heads in disbelief, because experiment after experiment is showing that the universe is entirely different from what astronomers had assumed since the beginning of modern science. Ordinary matter is the exception, and unknown, exotic matter is the norm. Our universe is mostly dark, and most of that dark matter is unknown, ineffable stuff that has never been seen directly. Had there not been so many experiments forcing cosmologists to accept this picture, it would seem utterly ridiculous. Chicago cosmologist Michael Turner asks, incredulously, "Who ordered this?"

Yet there is new hope of unraveling this matter mystery. Using enormous machines designed to re-create conditions in the first microseconds after the big bang, physicists are probing the origins of matter. They hope to understand matter's secrets by going back to its birthplace.

Chapter 8
The Big Bang in Our Backyard
[THE BIRTH OF BARYONS]

Three quarks for Muster Mark!

Sure he hasn't got much of a bark

And sure any he has it's all beside the mark.

— JAMES JOYCE, *FINNEGANS WAKE*

Not since the big bang had matter been in such a state. For a few microseconds after the birth of the universe, quarks and gluons roamed free in a blazing hot jumble of matter: a quark-gluon plasma. But the plasma quickly cooled, and the quarks and gluons formed more familiar particles, like protons and neutrons. The quark-gluon plasma condensed and disappeared. The universe was simply too cool for the quark-gluon plasma, just as the surface of the Earth is too cool for a puddle of molten iron.

Scientists have now brought us back almost to the moment of the big bang. Huge colliders have begun to re-create the conditions of the first few microseconds of the universe. By accelerating heavy nuclei to 99.99 percent of the speed of

light and slamming them together, physicists pour so much
energy into so tiny a space that they are creating tiny big
bangs. There are already hints that they have made a quark-
gluon plasma. Fourteen billion years after the plasma ceased
to exist, a laboratory on Long Island, New York, has revisited
the first few microseconds after the birth of the universe.

Though a handful of protesters tried to stop the experiment, fearing that the tiny explosions would destroy the universe, the scientists forged on, in hopes of seeing the first
moments of creation and revealing the origin and nature of
matter.

To understand how matter came about, we will have to dig
deeper and deeper into its nature. In a sense, we are taking a
trip backward in time. Right now, the universe is made up of
atoms. Strip off the electrons and look at the nuclei, and you
are back to an era 400,000 years after the big bang. Go even
further back—add more and more energy to your atoms—
and you can split them apart or smash them together and
make them stick. The center of a hydrogen bomb is as hot as
the first few minutes after the big bang.

Nevertheless, a few minutes after the big bang is not
enough. If you want to unravel the mysteries of matter, you
have to get hotter still. Add still more energy and even the
very protons and neutrons that make up an atom shiver apart
into their component quarks. This is the domain of particle
accelerators. As scientists build bigger and more powerful
accelerators, they reach closer and closer to the moment of
creation. Scientists are now beginning to see the first few microseconds of the universe. They hope to see a clue to the origin of matter.

The language of particle physics is somewhat bewildering at first, and it can seem like a scientist's fantasy, but the
theory behind it is incredibly powerful. Just as the periodic
table of elements describes the way ordinary matter behaves,
the *standard model* of particle physics describes the way all

known matter behaves. It's a tough language to pick up, but it is the language that nature seems to use.

The story begins with ordinary, baryonic matter. Ordinary matter is all around us, but it is not nearly as simple as it appears. Gases, liquids, and even solids like the book you are holding are mostly empty space. For about a century, scientists have known that the atom, the fundamental unit of everyday matter, is made up of a collection of smaller particles separated by vast amounts of emptiness. The center of an atom, its nucleus, is made up of two very similar particles: the neutron and the proton. They weigh almost exactly the same; the only obvious difference between the two is that the proton carries an electric charge and the neutron does not. The two are so similar that a neutron, left to itself for a few minutes, will spontaneously change itself into a proton by ejecting an electron (and another particle that we will meet later).

Bound to the nucleus by mutual electric attraction, the electron is a very different type of particle from the proton and the neutron. It is much less massive; it weighs about a two-thousandth as much as the proton and neutron, and carries an electric charge equal and opposite to the proton's. In modern terminology, the electron is a lepton, so named because it is light; the neutron and proton are known as baryons because they are heavy. For a few decades, chemists and physicists thought that this was the end of the story. Matter was made up of protons, neutrons, and electrons. By studying the interactions of these three types of particles, scientists could understand all the matter in the universe. But it turns out that there is much more to the story of matter. There are many more beasties in the particle zoo.

Over the years, physicists have discovered a whole menagerie of baryons like the proton, leptons like the electron, and middleweight mesons like the pion. For instance, physicists now know that the electron has two heavier siblings with very similar properties: the muon (symbolized by μ, the Greek lowercase letter mu), which is 200 times as massive as the electron, and the tau particle (symbolized by τ, the Greek lowercase letter tau), which is more than 3,500 times as massive.[1] The electron is the most common of the trio, the muon is the next most common, and the tau particle is the rarest, by far. The number of particles in the universe seemed to multiply like rabbits. It was a confusing picture, with all these leptons, mesons, and baryons floating about.

If this is not complicated enough, there is also the matter of antimatter. Antimatter might seem to be the fanciful creation of a science fiction author, but it actually exists. Scientists have been experimenting with it for more than seventy years.

In 1928, physicist P. A. M. Dirac created an equation that seemed to explain some of the more mysterious properties of electrons, but in the process his equation predicted something quite unexpected. If one took Dirac's equation at face value, there had to be an equal and opposite twin of the electron. It would weigh the same amount, have a charge of $+1$ instead of -1, and if it ever came into contact with an electron, the two would annihilate each other in a flash of energy.[2] This idea— antimatter—was so ridiculous that another founder of quantum theory, Werner Heisenberg, called Dirac's theory "the saddest chapter of modern physics." Nevertheless, in 1932 American physicist Carl Anderson, studying the smoky tracks that particles left behind as they passed through a chamber full of vapor, discovered the electron's antimatter twin: the positron, also known as the antielectron. Dirac was right. The electron has an antimatter twin.

1. Of course, this leads to the perverse state of affairs in which a lepton, the tau particle, is heavier than a baryon, the proton. The reason for sticking with this terminology, even after the heavy/light particle distinction gets blurred, will become clear.
2. Einstein's famous equation, $E = mc^2$, showed that matter can be converted into energy and vice versa. This is why particle physicists tend to express the mass of particles not in kilograms, a unit of mass, but in electron volts, a unit of energy. (A single electron volt is the amount of energy an electron would pick up while it flies between metal plates that are attached to a one-volt battery.) An electron's "mass" is 0.511 million electron volts (MeV), as is the antielectron's, so the flash of energy would total a little more than one MeV, all told. While this is a lot of energy on the subatomic scale, it would take about 250 *trillion* MeV to light a forty-watt bulb for one second.

Every massive particle, as it turns out, has an antimatter doppelgänger. The proton has an antiproton, the neutron its antineutron, and so forth.[3] Dozens of baryons, mesons, leptons, and their antimatter partners formed an ever growing menagerie. Making matters worse, they could change identities. If they are smashed together, they can change into entirely different particles. Leptons, mesons, baryons, antileptons, antimesons, antibaryons were multiplying, decaying, and changing into one another. Scientists could make little sense of the increasingly complicated particle zoo until the mid-1960s.

This is where quarks come in. Just as the introduction of the periodic table transformed a miscellaneous mess of elements into a systematic, coherent picture of the chemical properties of atoms, Murray Gell-Mann's idea of quarks transformed the chaotic zoo of particles into an ordered, understandable structure. Instead of treating the proton, neutron, and other baryons as indivisible, fundamental particles, Gell-Mann realized that he could explain the whole zoo of baryons by assuming that they are made up of three smaller particles — quarks. For instance, Gell-Mann proposed that an *up* quark has a charge of $+2/3$ and the *down* quark has a charge of $-1/3$; that led him to explain the charge of a proton ($+1$ from two ups and a down) and the neutron (0 from two downs and an up).

In fact, a few simple rules can explain *all* the properties of the baryons in the vast mess of particles, at least all the baryons we know of. These rules may look as though they come out of nowhere, but they work incredibly well. (Don't worry, there won't be a quiz!)

Rule 1: There are six *flavors* of quarks: up, down, strange, charm, bottom, and top. Any quark has a *color:* red, green, or blue. (The term *color* should not be taken any more literally than the term *flavor*. It's just a convenient way to distinguish a property of quarks.)

---

3. When antimatter and matter come into contact, they annihilate each other, so if you ever run into your antimatter twin, whatever you do, don't shake hands!

Rule 2: If a red quark, a green quark, and a blue quark combine, the result is a *white* colorless particle, just as red light, green light, and blue light yield colorless white light when combined.

Rule 3: A particle with color can never be observed directly. Thus physicists will never see an individual quark, because it has to be red, green, or blue. But they *can* see white particles, because they do not have color. This is why baryons, like the protons, are always made up of three quarks; one of the quarks is red, one is blue, and one is green, canceling out the colors of the individual quarks.

Rule 4: Each quark has a *spin*. It's best to think of spin like the spinning of a top; just as a top can spin two ways, clockwise or counterclockwise, a particle with spin can have positive spin or negative spin. (A particle's spin affects some of its properties, but we won't get into that just yet.) A quark most commonly takes on a spin of $+1/2$ or $-1/2$, at least in the simplest mathematical framework for understanding these particles.[4]

Rule 5: Each quark has an antiquark, and the antiquarks come in antired, antiblue, and antigreen varieties and also have a positive or negative spin.

These seem like arbitrary rules, but if you look at the zoo of particles with these rules in mind, you can explain the properties of every single baryon, and all the mesons too. The baryons are all made up of three quarks—one red, one green, and one blue—because three quarks (or three antiquarks) are required to get a white, colorless particle. There's also another way to get a colorless particle: take, say, a blue quark

---

4. In truth, it seems that the quark has less spin than the naive mathematical framework would suggest, because the gluons, the particles that hold the quarks together, have spin as well.

and attach it to an antiblue antiquark. The blue cancels out the antiblue, leaving a colorless particle: a meson. All mesons are made up of a quark and an antiquark bound uneasily together.

Instead of having to catalog dozens of dozens of particles, each with its own properties, the theory of quarks, their colors, and the forces that bind them—known as quantum chromodynamics—reveals the underlying structure behind the collection of subatomic particles. And just as chemists predicted the existence of unknown elements by looking at empty spots in the periodic table, physicists predicted the existence of as-yet-undiscovered particles, such as the omega minus (Ω^-) baryon,which was discovered in 1964. For his insight, Gell-Mann won the Nobel Prize in 1969.

Even though scientists cannot see quarks directly, they are quite convinced that quarks exist; quantum chromodynamics is more than a mere mathematical formalism. When a physicist zaps a proton with an x-ray, the x-ray bounces off the proton as if it is made up of three smaller particles with charges of 2/3, 2/3, and −1/3, as the quark picture implies. Whether scientists use an x-ray or an electron to probe the inside of a proton, a neutron, or another baryon, the result is the same; the baryon appears to be a composite object, and the elements that make up the baryon have just the properties that quantum chromodynamics predicts.

There's yet another ingredient to quantum chromodynamics. A baryon, such as the proton, is made up of more than just quarks. Another type of particle holds the quarks together: the appropriately named *gluon*. Gluons are the glue that makes quarks stick together. These are the gluons of the quark-gluon plasma that filled the very early universe. Since the proton is so hard to split apart, this sticky force is incredibly strong. Physicists, in a rare outburst of creativity, dubbed this force the *strong force*.

The strong force is what keeps the quarks in baryons and mesons stuck together; it also binds the protons and neutrons to each other in a nucleus. Scientists believe all this without ever having seen an individual quark; after all, a quark, which has color, is always confined with other quarks to make a colorless particle. Because of the rules of the strong force, quarks cannot roam free—and they have been confined since the first few microseconds after the big bang. But in the early universe, this wasn't the case.

Like any force, the strong force can be defeated if you put enough energy into it. For instance, the force of gravity confines you to the surface of the planet. If you jump, you can escape the surface briefly, but you will eventually come crashing back down, because your legs aren't strong enough to defeat the gravitational attraction of the Earth. You are bound to the Earth by the gravitational force. Even powerful machines like jets are only able to overcome the effects of gravity for a short time; they must eventually come back to Earth. However, if you put an enormous amount of energy (and money!) into the effort—if you strap yourself inside a cramped chamber atop an enormous tower of explosives—you can escape the Earth's gravitational pull and fly toward the moon or out of the solar system entirely. You can defeat the force of gravity. Similarly, water molecules are loosely bound to one another by the electromagnetic forces between individual molecules.[5] Heat water enough, though, and the electromagnetic attractions are unable to keep the molecules bound to one another. As water boils, individual molecules, too energetic to be bound by the electromagnetic forces that hold the water together, fly away from the group, no longer confined to the liquid. Liquid water boils, evaporates, and becomes a gas. Put enough heat into a tub of water and you can defeat the electromagnetic force.

So too can the bonds of the strong force be defeated. If you

5. You can see this phenomenon directly; comb your hair to build up a static electrical charge on the comb, and then hold it near a tap where water is dribbling out. You will see the water stream bend toward the comb, attracted by the electrical charge on the comb.

smash two atoms together at high speeds, or if you slam them hard enough with other particles, you can break the strong-force bonds that make the protons and neutrons stick together. Atoms fly apart into smaller bits when the strong-force bond that keeps the nucleus in one piece is defeated. If the nucleus is heavy enough, you release a bit of energy when the breakup happens; this process, fission, is the process behind the atomic bomb. But even in the heart of an atomic bomb, there is not enough energy to break apart the individual protons and neutrons in the nuclei. Even as the strong force that holds the nucleus together is defeated, the strong force that keeps the quarks bound to one another in little packets—protons and neutrons—holds tight. It has held tight for fourteen billion years.

The binding force failed to hold only for a very, very brief moment after the birth of the cosmos. When the universe was extremely hot and dense, just a few fractions of a second after the big bang, every particle in the universe was imbued with so much energy that the force was just too weak to bind the quarks to one another. Quarks and gluons floated about, unconfined, in the quark-gluon plasma. As the universe expanded and cooled, the strong force got the upper hand. About a millionth of a second after the big bang, the quark-gluon plasma condensed into protons, neutrons, and other particles, just as water vapor, cooled by a pane of glass, condenses into droplets. Never again were quarks let out of their cages; they never appeared again without being bound to an antiquark or two other quarks by the strong force. At least, this was the case until a few years ago.

Scientists have re-created conditions only known in the first few microseconds after the big bang. They now believe that they have broken the bonds of the strong force and, for the first time since the birth of the universe, released quarks from their age-old confinement. This means that scientists are on the brink of understanding where all the baryonic matter in the universe came from.

The first hints of quark parole—what physicists term *deconfinement*—came from an accelerator buried under the earth near Geneva, Switzerland. The Super Proton Synchrotron machine is essentially a six-kilometer circle of magnets that smashes atoms together at enormous speeds—at more than 99 percent of the speed of light. At that speed, a lead nucleus, which ordinarily looks like a ball, is as flat as a dinner plate, thanks to the relativistic distortion of space and time. And when one of these dinner plates smacks into something, it explodes in spectacular fashion. So much energy is packed into so little space that on a very tiny scale the explosion looks like a tiny big bang. Under such conditions, even neutrons and protons can't hold together. They will evaporate in a puff of energy, freeing the quarks from their fourteen-billion-year-old confinement. However, the deconfinement can only last for about 10^{-23} seconds before the quarks and gluons created by the intense burst of energy condense into thousands of baryons and mesons that fly off in all directions. Scientists do not see the quark-gluon plasma directly; they see the tracks of particles that zoom off after the quark soup has refrozen into baryons and mesons. From those tracks, they try to infer the conditions of the explosion. It's something like trying to figure out how an automobile works by smashing two together and analyzing the trajectories of hubcaps, fenders, and other parts that fly off. It's a tricky, and controversial, job.

In February 2000, CERN, the European particle physics laboratory that runs the Super Proton Synchrotron, made a tentative announcement that its researchers might have finally seen a quark-gluon plasma. Their strongest evidence was based upon the curious reduction in a meson known as the J/ψ, or J/psi. (Pronounced "jay-sigh" because the second symbol is the Greek letter psi, the J/ψ meson got its awkward name because two competing groups discovered it at the same time; one called it the J particle, and the other called it ψ.) The J/psi is made up of a charm quark and a charm antiquark, and since the charm quark is relatively rare, so too is the J/psi.[6]

Particles like the J/psi can be made in energetic collisions; it's the reverse of what happens when matter and antimatter collide, annihilating each other and releasing a burst of energy. Put enough energy in a small space by smashing two speeding atoms together, for example, and you can create matter-antimatter pairs, such as a positron and an electron, or a charm quark and charm antiquark. (You can't take pure energy and make a charm quark without its antiquark; you have to make a matter-antimatter pair.) Ordinarily, a newborn charm quark and charm antiquark, created along with many other quarks and antiquarks in an energetic collision, would instantly pair up or triple up with nearby quarks and antiquarks, forming mesons or baryons. Since a charm quark is produced with a charm antiquark, it is rather likely that the charm quark and charm antiquark bind together and form the J/psi meson. But if there is a quark-gluon plasma, quarks don't immediately condense into mesons and baryons. They float about in a chaotic, seething soup of quarks and gluons and only condense when the temperature drops far enough. In such a soup, the relatively rare charm quark would probably drift away from its antimatter twin and, more than likely, would pair up with the more common up or down quark rather than with a charm antiquark. Thus, it would become much less likely to form a J/psi meson.

In the 1980s, scientists predicted that one of the signs of a quark-gluon plasma would be a relative scarcity of J/psi particles. This is precisely what CERN claimed to have seen, as well as a weak indication of other signatures of the quark-

6. In general, the heavier a particle or quark is, the rarer it is, for reasons will be explored in chapter 12. Up and down are, by far, the lightest and most common quarks; strange is the next most common. Charm comes next, then bottom, then top. A charm quark weighs roughly a thousand times as much as an up quark, so it is considerably harder to find and is harder to make in a collider. This mass/rarity relation is also the reason that the tau particle is so much rarer than the muon, and the muon is so much rarer than the electron.

gluon plasma. However, they did not come right out and claim to have created the quark-gluon plasma.

"A common assessment of the collected data leads us to conclude that we now have compelling evidence that a new state of matter has indeed been created," said CERN physicist Maurice Jacob. That new state of matter "features many of the characteristics of the theoretically-predicted quark-gluon plasma." In other words, it looks like a duck and walks like a duck.

Other scientists weren't so sure that it was, in fact, a duck. William Zajc, from Columbia University, argued that CERN's evidence wasn't sufficient to prove that its investigators had seen a quark-gluon plasma. "It's fair to say I don't find they've made a compelling argument for the discovery at all," he said, shortly after CERN presented its evidence. At the time, Zajc was busy getting ready for the startup of a more powerful accelerator, based at the Brookhaven National Laboratory in New York. This accelerator, the Relativistic Heavy Ion Collider (RHIC, pronounced "rick"), was five times more powerful than the Super Proton Synchrotron—and was expected to have the power to create the quark-gluon plasma. Sure enough, as soon as it was turned on, scientists began to see even more conclusive evidence that they had freed quarks from their color confinement.

RHIC's magnets are so powerful that some hysterical protesters feared that scientists using them would inadvertently end the universe. When the accelerator slammed a gold atom into a gold atom with sufficient force, they argued, a chain reaction would begin whereby the current universe would become unstable—and at the speed of light, an expanding wave of destruction would wipe away all the matter in the universe, leaving nothing but energy in its wake. It's a far-fetched idea, but there is a tiny grain of truth beneath the fear. Luckily, as most serious scientists knew, the worries were greatly overblown.[7]

---

7. Extremely high-energy cosmic rays slam into the moon all the time, so from those interactions, which pack as much punch as a RHIC collision, you can show that the universe should have ended a long time ago if the doomsayers' argument were correct.

By November 2000, RHIC's first results began to appear, and they soon showed the subtle signature of deconfinement. At low energies, a nucleus behaves something like a clump of hard wax pellets. Slam two into each other, and particles shoot in all directions, caroming off one another like hard billiard balls. But in RHIC's superpowerful collisions, something different happens. Though scientists see violent jets of particles coming off the sides of the collisions, they see fewer of these high-energy jets than expected. Just as someone counting wax pellets might explain that the wax had melted at high energies, creating a sticky mess, particle physicists suspects that the particles in the nuclei might be melting into a sticky quark-gluon plasma. Particles passing through the plasma are forced to slow down, and the jets of particles shooting out the sides lose some of their energy. The jets are quenched, and RHIC was the first experiment to see evidence for jet quenching.

There are other lines of evidence for a quark-gluon plasma. When two of these dinner-plate gold atoms smack into each other, they usually don't hit each other directly face on. Instead, the nuclei strike each other off center, colliding only in an almond-shaped region where the disks overlap. (Think of two dinner plates hitting each other slightly off center. The region where they actually contact each other is the region in question.) If the protons and neutrons stay intact, the particles should explode outward, roughly evenly distributed in all directions, destroying the almond-shaped pattern of the collision. However, scientists were surprised to see that the almond-shaped distribution was visible in the way that the particles sprayed away from the collisions. Scientists have difficulty explaining this behavior unless they assume that the protons and neutrons break apart, forming a quark-gluon plasma. When this happens, the particles spread their energy more evenly among the individual quarks in the almond-shaped collision region, and after they recondense, they spray away in an almond-shaped pattern. It is not conclusive evi-

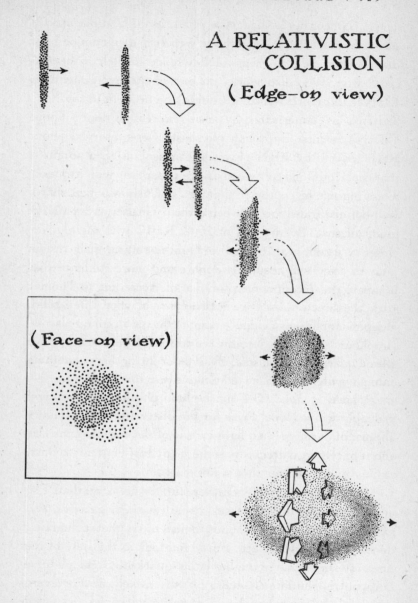

A RELATIVISTIC COLLISION

(Edge-on view)

(Face-on view)

A heavy-ion collision at more than 99 percent of the speed of light

dence, but it is fairly suggestive of a quark-gluon plasma. "It seems to imply that something weird is happening," says James Thomas, a physicist at Lawrence Berkeley National Laboratory who also works on one of the experiments at RHIC. "But more than that wouldn't be prudent to say."

Just as American physicists expressed doubt about CERN's results, European physicists were reserved about RHIC's. While CERN physicist Carlos Lourenço admitted that the RHIC measurements were consistent with a quark-gluon plasma, he cautioned that they don't show a nice, sharp, well-defined transition between ordinary matter and a quark-gluon plasma. By the end of 2002, RHIC was edging ever closer to making an official announcement; as their results pour in, they will accumulate more and more evidence that points to the deconfinement of quarks. According to Thomas Kirk, the director of science at Brookhaven, when three, solid, independent lines of evidence point to the quark-gluon plasma, Brookhaven will reveal to the world that its researchers have seen a quark-gluon plasma. By 2004 or 2005, there should be enough evidence to make a definitive case for discovery.

A lot is at stake. The quark-gluon plasma discovery will probably win a Nobel Prize for the discoverers. Much more important, it might shed light on one of the enduring puzzles about baryonic matter: why is there any matter at all? In fact, at first glance, matter should not exist.

This might seem like a rather stupid statement; there's so much matter in the universe—countless stars, galaxies, and galaxy clusters—that it sounds foolish to say that it shouldn't be there in the first place. But if you look at the state of the universe shortly after the big bang, quarks and gluons were born out of the intense energy of the cataclysm. For every quark, there is an equal and opposite antiquark, so when three quarks gather together to form a baryon, in all probability, somewhere else in the universe, three antiquarks are condensing to make an antibaryon. When the universe was very young, there should have been roughly the same amount

of matter and antimatter in the universe. Matter and antimatter annihilate each other, so equal parts of matter and
antimatter should have completely wiped each other out, annihilating each other particle by particle. There should be
nothing left.

However, there is plenty of matter in the cosmos, and as
far as scientists can tell, very little antimatter. This seems to
imply that there was more matter than antimatter in the early
universe. When the matter and antimatter annihilated each
other, some matter was left over. This leftover matter, just a
fraction of what the universe originally contained, makes up
all the stars and galaxies, and all the baryonic dark matter in
the cosmos. For some reason, whatever process created the
baryonic matter in the universe must have preferred matter
over antimatter. Matter and antimatter are not truly equal
and opposite: some subtle difference between the two made
matter easier to create than antimatter. The two aren't precise
mirror images; the symmetry between them is broken. To that
broken symmetry, we owe the existence of the matter in the
universe—and our own existence.

By taking a trip back to the beginning of the universe in a
particle accelerator, scientists can see what conditions led to
the victory of matter and the annihilation of antimatter. For
instance, RHIC physicists have already seen the ratio of antiprotons to protons produced in collisions rise from almost
zero (no antimatter produced) in collisions with the lowest
energies to about 65 percent (two antiprotons produced for
every proton created) in collisions with high energy. As the
RHIC accelerator gets more and more powerful, the conditions will get more and more like the newborn universe in the
moments after the big bang.

While scientists are awaiting the official announcement
that RHIC has re-created the conditions of the early universe,
other facilities are trying to probe the nature of matter and
antimatter by other means. After all, there's more to matter
than quarks and gluons.

Chapter 9
The Good Nus

[THE EXOTIC NEUTRINO]

Neutrinos, they are very small.

They have no charge and have no mass

And do not interact at all.

The earth is just a silly ball

To them, through which they simply pass,

Like dustmaids down a drafty hall

Or photons through a sheet of glass. . . .

—JOHN UPDIKE, "COSMIC GALL"

Even as scientists close in on the birth of baryons, of ordinary matter, they realize that they are missing most of the story—the tale of nonbaryonic matter. All the experiments that weigh the mass in the universe agree. Baryonic matter is about 5 percent of Ω, but the total amount of matter in the universe is about seven times that. This obviously means that the vast majority of the matter in the universe is unaccounted for. Most of the matter is missing in action. The remainder must be a form of matter that is nonbaryonic, something "exotic" that is not made of quarks. Luckily, there are other particles just as fundamental and indivisible as quarks. The electron is the most familiar of them, and scientists also have a good understanding of the electron's heavier siblings, the muon and the more massive tau particle. These three particles make up half of the cadre of leptons.

All the matter in the universe—all the matter that scientists have encountered, anyhow—is made up either of quarks or of leptons. The exotic matter in the universe cannot be made up of quarks, so we must turn our attention to the leptons to solve the mystery of the missing matter. However, the three leptons we have met so far cannot account for all the exotic matter in the universe. Muons and tau particles are unstable; the muon decays in a millionth of a second, the tau in less than a trillionth of a second, so they can't hang around long enough to account for much of the missing matter.[1] Even the electron, as stable (and as common) as it is, cannot account for that much of the exotic matter; the charge it carries would give it away if it were floating about in space, and scientists have no indication that an enormous number of unbound electrons are zooming about the universe. By the process of elimination, suspicion naturally falls upon the three remaining leptons: the neutrinos. But neutrinos are, by far, the most misunderstood and elusive of all the particles in the menagerie. Until a few years ago, nobody knew how much the neutrinos weighed, or even if they weighed anything at all. The neutrino was almost undetectable, so scientists could not measure even its most basic properties. Nobody knew whether the neutrinos had mass or whether they traveled at the speed of light.

In the past few years, the fog that surrounds the neutri-

---

1. Even though the leptons are fundamental particles, they can decay into other, more stable particles by processes that will be described shortly. Physicist Gordon Kane at the University of Michigan makes the important point that the term *decay* is misleading when it comes to particle physics. As he writes in his book *Supersymmetry*, "A major difference between the way *decay* is used in physics and its use in everyday life is that the particles that characterize the final state are not in any sense already in the decaying particle. The initial particle really disappears, and the final particles appear."

nos has finally begun to clear. Scientists are weighing the neutrino and cataloging its properties. Neutrino astronomers are even using them to analyze objects in the sky, just as ordinary astronomers use particles of light, photons, to do the same thing. The age of the neutrino has arrived. With its arrival, cosmologists are finally beginning to learn about the mysterious exotic substance that, totaled up, outweighs the baryonic matter in the universe as dramatically as an automobile outweighs a person.

The first half of the story of mass had to do with quarks and gluons, the creatures of the strong force. The second half of this story switches from quarks to leptons and from the strong force to the *weak* force. These are the final components in the standard model, the overarching theory that guides scientists in their quest to understand the matter in the universe—and where all the missing matter is. Before cosmologists can understand the vastness of the universe, the particle physicists must teach them what the cosmos consists of. Right now, the hottest topic in particle physics is the neutrino.

At first, the neutrino was simply an accounting device. In 1930, physicist Wolfgang Pauli realized that Nature seemed to be fudging her books during a process called beta decay. We've already briefly encountered beta decay; it's the process whereby a neutron turns into a proton. Beta decay happens all the time in nature; not only do free neutrons decay in this way, but certain unstable elements, like cobalt-60, become more stable by emitting an electron and turning a neutron into a proton. Pauli realized that something was wrong when beta decay seemed to be violating one of the most fundamental rules of physics: momentum was not conserved.

A few ideas in physics are almost sacrosanct and seem to be fundamental to the way the universe works. For instance, scientists believe that energy can be neither created nor destroyed; it can dissipate, it can change forms, it can turn into matter, but it can't disappear entirely and it can't come out of nowhere.[2] This is the law of energy conservation: the amount of energy that exists before any event must be the same as the amount of energy that exists after the event. Nature seems to keep her books very carefully when it comes to energy.

The same is true with a quantity called momentum. It is easiest to think about momentum as something like "pushing power," and it tends to be a function of an object's mass and its speed.[3] The larger an object and the faster it moves, the more momentum it has—and the less you would like it if it slams into you. For example, a car moving at fifteen miles an hour will do more damage when it crashes than one moving at five miles an hour; it has more momentum because it moves faster. A bus moving at fifteen miles an hour is more dangerous still, because it has more mass, and thus more momentum.

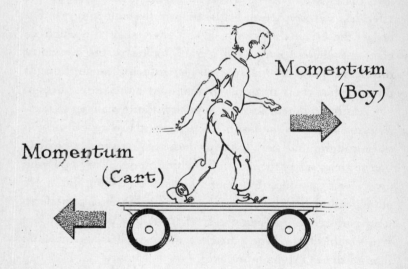

Conservation of momentum

2. Although it might seem like it when it comes to vacuum energy—more on this shortly.
3. Some objects, like photons, can have momentum without having mass, but ignore that for the moment.

Just like energy, momentum can be transferred or redistributed, but in the absence of an external force it cannot be created or destroyed. You can see this principle in action if you have a child's toy wagon or a wheeled dolly handy. Stand (carefully) atop the cart at one end and then walk to the other end. You'll notice that as you walk forward, the cart moves backward, rather than staying put. This is because you and the cart start off with no momentum, and when you start walking, you acquire some momentum because you are moving faster. To counteract this effect, the cart picks up some negative momentum; it moves in the opposite direction, maintaining a net momentum of zero.[4]

The process of beta decay has a lot in common with the toy wagon example. A stationary neutron, with zero momentum, spits out an electron, which has momentum. To counteract the effect, the neutron acquires a bit of negative momentum; it flies off in the other direction—more slowly than the electron, since it has more mass. When Pauli looked at the motion of the electron and compared it to the motion of the neutron, he realized that the particles' momenta did not quite add up to zero. Some momentum was left over. Pauli realized that either the law of momentum conservation had to be thrown out, or some invisible, (nearly) massless, no-see-um particle had to carry away that extra little bit of momentum by flying away from the decaying neutron. Pauli chose the latter. He concluded that there had to be an invisible little particle to help Nature keep her books straight. A few years later, Enrico Fermi dubbed the particle a neutrino, Italian for "little neutral one." The name stuck.[5]

How could scientists prove that Pauli's creation was not a mere figment of his imagination? It is incredibly difficult to detect these invisible particles. They have little or no mass,

---

4. The author accepts no liability for injuries sustained while performing momentum experiments.
5. Technically, what zooms off in a beta decay is an antineutrino. Except in a context in which it's important to distinguish neutrinos and antineutrinos, physicists tend to refer to both as neutrinos.

making them almost immune to the pull of gravity (though there are a huge number of them, making them important on cosmic scales); they have no charge, making them almost indifferent to electric and magnetic forces; and they do not even feel the tug of the strong force that binds quarks so tightly together. Because they are not deflected or affected by these forces, it is almost impossible to capture one or detect its presence. In fact, neutrinos (and antineutrinos) are so scornful of their surroundings that they seldom condescend to interact with matter—such as the matter that makes up any sort of detector—at all. A typical neutrino would pass through the entire Earth without noticing the enormous lump of matter it was passing through.

However, scientists spotted the tiny particles in 1956, when Frederick Reines and Clyde Cowan, physicists from the Los Alamos National Laboratory, detected antineutrinos pouring out of the fission reactor at the Savannah River nuclear plant in South Carolina. According to theory, the nuclear reactions were churning out antineutrinos at an enormous rate, so Reines and Cowan placed more than a ton of liquid in a room that was supposedly full of them. The liquid target was full of protons, and when it was bombarded with antineutrinos, the physicists saw the reverse of beta decay. (If an antineutrino strikes a proton, the proton spits out an antielectron, turning itself into a neutron.) Despite the huge number of antineutrinos passing through such an enormous target, an inverse-beta event was fairly rare, happening only every other minute or so. But as rare as these events were, they were sure signs that antineutrinos exist; they were turning protons into neutrons. As indifferent as neutrinos and antineutrinos are to the pull of matter, Reines and Cowan showed that, once in a great while, one would condescend to interact with a target. (Reines received the 1995 Nobel Prize for the discovery.)[6]

6. Cowan had died two decades earlier, and the Nobel committee does not award posthumous prizes.

If neutrinos scarcely feel the effects of gravity, electric fields, and the strong force, how can they interact with matter at all? There is a fourth force that *does* affect neutrinos; it is not as powerful as the strong force, and it has an influence only over tiny distances. However, neutrinos do feel this so-called weak force, and it is thanks to the weak force that scientists have detected them.

The weak force is part of a mathematical framework that physicists have constructed over the years to explain the sub-atomic world. This framework, the standard model, describes the constituents of matter—the quarks and leptons—and the interactions between them, which are a result of the forces: the strong, weak, and electromagnetic forces.[7] (Gravity and particle masses are not directly included in the standard model, for reasons that will be discussed later.) The standard model is incredibly successful at predicting the properties of matter—astonishingly so. For example, physicist Hans Dehmelt won the 1989 Nobel Prize for measuring how electrons twist under the influence of a magnetic field. The value he measured matched the value predicted by the standard model to ten decimal places. By any account, the standard model did a spectacular job, and it is for this reason that physicists are so attached to it.

The forces in the standard model are carried, or *mediated*, by particles that interact with quarks and leptons. We have already encountered two of these force carriers. The gluon,

---

7. In truth, the electromagnetic force and the weak force have been "unified"—that is, they have been shown to be different facets of the same underlying phenomenon, even though they look completely different. Just as a glass vase and a pile of sand appear to have entirely different properties, if you heat them both to a high enough temperature, you can see that they are actually the same substance, silicon dioxide. Analogously, the electromagnetic force and the weak force are actually the same underlying force, which becomes obvious at extremely high temperatures. (Sheldon Glashow, Abdus Salam, and Steven Weinberg showed that and received the 1979 Nobel Prize for this insight.) Scientists believe that the strong force, at still higher temperatures, will unify with this electroweak force, and they hope that they can include gravity as well in this grand unifying vision. If they succeed, they will have finally created the "theory of everything," the ultimate theory of how matter in the universe behaves.

symbolized by g, mediates the strong force. It is responsible for the attraction between quarks and for their color-based confinement. The carrier of the electromagnetic force—what makes electrons attracted to protons and what makes a refrigerator magnet attracted to steel—is actually the photon, symbolized by γ, the Greek lowercase letter gamma. Though physicists cannot actually see the photons zooming from proton to electron, they can see the particles behave as if photons are being emitted and absorbed.[8]

The weak force, though it is part of the standard model, is different from the other forces. Unlike the strong force or electromagnetism, it has the ability to change the flavor of quarks and leptons. For instance, it can turn a down quark into an up quark or a neutrino into an electron (or vice versa). For instance, in the decay of the neutron, the weak force turns a down quark into an up quark. Just as electromagnetic repulsion is mediated by photons, this decay is mediated by a carrier of the weak force; in this case, it is the W^- boson. (There are two other carriers of the weak force: the W^+ and the Z boson.)[9]

Neutrinos, symbolized by the lowercase Greek letter nu, ν, are pretty much indifferent to the effects of photons and gluons, but they feel the influence of both types of W boson and the Z—the weak force—so scientists can detect the presence of a neutrino when it has a weak interaction with matter in a detector. For instance, Reines and Cowan detected an antineutrino when it exchanged a W particle with a proton, turning the proton into a neutron.

8. Physicists call these photons *virtual,* and they have a different sort of presence from that of a "real" photon. It is a mindboggling concept that we will discuss when we get to the particles that fill up the vacuum.
9. The Z boson was predicted by the theory of electroweak interactions before it was spotted, another indication that the theory truly captures the essence of particle interactions. It is possible that there are other, undiscovered weak-force carriers, such as the so-called Z-prime. Incidentally, the terms *boson* and *fermion,* which will become significant later, refer to the amount of spin that a particle has. Bosons and fermions have radically different properties. All the leptons and quarks are fermions, while all the force carriers are bosons.

Once scientists figured out how to detect the presence of neutrinos, they gingerly began to figure out their properties. The first surprise came in 1962, when Leon Lederman and his colleagues at Columbia University and the Brookhaven National Laboratory discovered that there was more than one type of neutrino. They were studying the decay of certain mesons into one of the heavy siblings of the electron, the muon. When Lederman's team detected the neutrinos (technically, antineutrinos) produced by the reaction, they noticed something odd. The neutrinos involved in this muon-creating reaction produced only muons when they interacted with matter. The neutrinos involved in beta decay, on the other hand, produced only electrons. There seemed to be two different types of neutrinos with different behaviors. One type was associated with electron reactions, and another with muon reactions. It was a surprising discovery: there was more than one flavor of neutrino. The invisible particles came in different varieties. The ones that interact with electrons, electron neutrinos, acquired the symbol ν_e; muon neutrinos, ν_μ; tau neutrinos — which were only directly detected in 2000 because of the rarity of the tau particle that it interacts with — quite naturally got the label ν_τ. (Lederman received the Nobel Prize in 1988 for the discovery of neutrino flavors.)

These three neutrinos are the last pieces of the standard model puzzle. Previous chapters described the six flavors of quarks: up, down, strange, charm, bottom, and top. Now add to that the six leptons: the electron, muon, tau, and the associated neutrinos. These are all the fundamental particles that make up matter. The force carriers, the particles that make matter interact with matter, are the photon, the gluon, the two W bosons, and the Z. The interactions between these particles and the force carriers describe almost all the fundamental properties of matter. The standard model contains them all; it is a stunningly powerful theory, and it has been one of the great successes of the twentieth century.

Naturally, with such a powerful theory, cosmologists

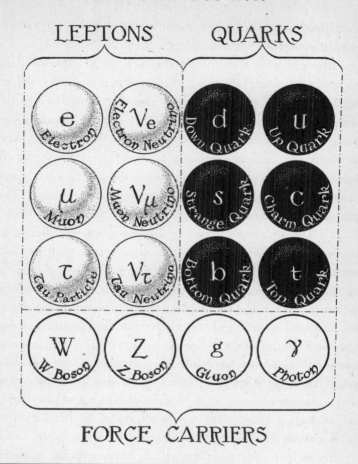

The fundamental particles of the standard model

thought they could explain the missing matter in the universe—the nonbaryonic stuff that makes up six-sevenths of the mass in the cosmos—by looking at the particles in the standard model. The quarks are not candidates, as they make up baryonic matter. The electron, muon, and tau particle aren't candidates, because they are charged, and the muon and tau particle decay too quickly. That leaves the neutrinos as candidates for the missing dark matter. But how much mass do they have? Are they heavy enough to outweigh the baryonic matter in the universe? Scientists are just finding out. In the process, they are stretching the standard model to its limits.

Taken literally, the plain-vanilla form of the standard model does not say anything about particle masses at all; in fact, if theorists try to put mass into the equations of the model, the equations blow up and become all but meaningless.[10] The standard model is agnostic about how much mass the neutrinos have; neutrinos can even have no mass at all. Indeed, physicists preferred a massless neutrino, because something rather weird happens if neutrinos have mass. They spontaneously change flavors.

If an electron neutrino, say, has no mass, it will always stay as an electron neutrino until it interacts with another particle—it would behave as an electron does, always keeping its identity. But once physicists assume that the neutrino has mass, by forcing some extra terms into the equations of the standard model and cleaning up a mathematical mess, then the equations that describe the neutrinos smear into one another. An electron neutrino can no longer be considered a "pure" electron neutrino; rather, according to those equations, it acquires some of the nature of muon and tau neutrinos as well.[11] (Such things often happen in the quantum world.) Also, the degree of smearing does not stay the same; as an electron neutrino travels, it imperceptibly acquires more and more of the muon neutrino's nature, until it actually becomes a muon neutrino. The electron neutrino must *oscillate* into a muon neutrino and a tau neutrino, and vice versa.

10. Physicists solve this problem by extending the standard model to include a new particle known as the Higgs boson, which imbues particles with mass. The Higgs is likely to be discovered in the next decade; if it isn't, particle physics is in big trouble. The discovery of the Higgs will be a tremendous accomplishment for the understanding of the physical world, but since it doesn't have a direct consequence to the field of cosmology, Higgs physics is beyond the scope of this book.

11. Technically, the three neutrinos are a mixture of three *basis elements*, known, confusingly, as v_1, v_2, and v_3. An electron neutrino, for example, might be mostly v_1, with a little v_2 and v_3 thrown in for good measure, while a muon neutrino might be mostly v_3 with some v_1 and v_2 as well. The electron, tau, and muon neutrinos, therefore, are the same things blended in slightly different ways.

Thus, if you assume neutrinos have mass, their lives become much more complicated: they oscillate constantly, switching flavors as they travel. Conversely, if they oscillate, they must have mass.

Scientists naturally preferred the less-complicated picture. They assumed that neutrinos did not have mass, and that they do not oscillate. Since neutrinos didn't have any mass, cosmologists could safely ignore them; even though the universe is probably flooded with neutrinos, the massless particle couldn't contribute anything to Ω, because, without mass, a neutrino can't affect the curvature of spacetime or the amount of mass in the universe. And so things stood for more than three decades. The darn things were too difficult to detect; scientists couldn't look for oscillations or do any sophisticated tests of the neutrinos' properties.

In the last few years, the scientists' picture of the neutrino has suddenly and dramatically changed. A new generation of neutrino detectors has finally given physicists the ability to unravel the mystery of the neutrino, and they discovered that the neutrino has mass after all. This discovery has made cosmologists sit up and take notice: since the neutrino has mass, it holds some of the secret of the missing mass, the as-yet-undetected exotic matter that vastly outweighs the ordinary, baryonic matter in the cosmos.

Deep below the mountains in Kamioka, Japan, a gigantic cylinder of water awaits the arrival of neutrinos. Studded with thousands of sensors to detect the interaction of a neutrino with a particle in the cylinder, the Super-Kamiokande neutrino observatory provided the first solid evidence that neutrinos have mass.

The Super-Kamiokande detector, known as Super-K, is really a sensitive light meter. When a neutrino reacts with a particle in the cylinder, an electron or muon (or tau, for that matter) flies off at great speed. In fact, it is flying so fast that it emits the electromagnetic equivalent of a sonic boom, a flash of light known as Cerenkov radiation. The cylinder's walls are studded with photodetectors to pick up that flash. Since lots of other energetic particles, like gamma rays or cosmic rays, can send a particle flying and make it flash, the Super-K detector has to be shielded from external sources of radiation, hence the necessity to bury the laboratory under an entire mountain of rock. Cosmic rays and gamma rays cannot get through the enormous shield; they are absorbed, because they interact readily with matter. Neutrinos, on the other hand, can pass through the mountain fairly easily and get picked up by the detector. Super-K was not the first experiment to be buried deep underground. However, because of its enormous size, Super-K is much more sensitive than its predecessors, and it also has the advantage of being able to tell, quite precisely, from which direction a given neutrino has come. This gave the scientists the ability to do a very interesting experiment involving atmospheric neutrinos, an experiment that gave scientists the first real understanding of the neutrino's nature.

The Earth is constantly bombarded by cosmic rays, particles with extremely high energies that smack into the atmosphere.[12] When a cosmic ray hits, it creates a shower of secondary particles, including neutrinos. These atmospheric neutrinos then proceed on their merry way. Super-K's great insight came when it compared the atmospheric neutrinos coming from above the laboratory with atmospheric neutrinos that came from below—which had formed on the other side of the Earth and then passed through the entire planet before hitting the detector. All these atmospheric neutrinos should look the same, because they are formed by the same processes—unless they oscillate. If neutrinos oscillate as they travel, then the neutrino streams that zoom through the entire Earth should change flavors somewhat as they travel. By the

12. These cosmic rays often have an enormous amount of energy. In October 1991, a Utah-based observatory spotted a cosmic ray—a subatomic particle—that packed as much energy as an 88-kilometer-per-hour baseball. They dubbed it the "Oh-my-God" particle.

time they reach the detector, the long-distance neutrinos should have different proportions of electron, muon, and tau neutrinos than the ones that travel only a few miles from the edge of the atmosphere to the detector. In other words, if the neutrinos from below are a different mixture of flavors from the ones coming from above, then neutrinos must be oscillating. In 1998, this is precisely what the Super-K team announced. They saw evidence for oscillations. This was an amazing discovery. For the first time, scientists realized that neutrinos must have mass, even if only a tiny amount.

In 2001, a second neutrino detector, buried two kilometers down a nickel mine in Sudbury, Ontario, clinched the case for neutrino oscillation, this time by looking at neutrinos

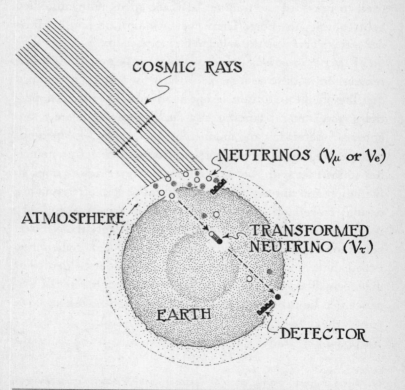

Neutrinos created in the atmosphere change flavor as they pass through the Earth.

from the sun rather than from the Earth. The sun is powered by fusion: hydrogen and other light elements in the sun crash together, forming a heavier element (like helium) and releasing energy in the process. Though the hydrogen-to-helium reaction is, by far, the dominant one, other nuclear reactions are going on in the sun, and traces of other elements are present besides hydrogen and helium. Boron-8 is particularly interesting since, as it beta-decays, it spits out a high-energy electron neutrino, which, almost indifferent to the effects of gravity and electromagnetic radiation, flies off into space. Many of these neutrinos pass through the Earth. Theorists thought they had a firm handle on how many neutrinos should be coming from the sun. However, when scientists tried to measure the number of these solar neutrinos, they kept coming up short. There were simply too few of these neutrinos. This was the solar neutrino paradox.[13]

That paradox was definitively solved in mid-2001, when researchers at the Sudbury Neutrino Observatory announced that they had found the missing neutrinos. The electron neutrinos streaming from the sun had simply oscillated into harder-to-detect tau and muon neutrinos and had escaped notice. (There are plenty of electrons around for electron neutrinos to interact with, but few muons and even fewer taus, so the muon and tau neutrinos interact much less often and are seldom picked up by the detector.) The sensitive Sudbury Neutrino Observatory was able to spot not only the electron neutrinos but the muon and tau neutrinos too, and when they added them up, they got the expected number of neutrinos that were flowing from the sun.[14] It was another bit of evidence that neutrinos oscillate, and that they have mass.

13. In 2002, physicists Raymond Davis Jr. of the University of Pennsylvania and Masatoshi Koshiba of the University of Tokyo won the Nobel Prize for their detection of neutrinos from the sun and from other astrophysical sources, which led scientists to the solar neutrino paradox.
14. The electron neutrino interacts with electrons via the W bosons, and it can interact with matter via the Z boson as well. The tau and muon neutrinos also can do the Z boson interactions, but due to the rarity of taus and muons, the W boson interactions are largely denied to them, making them harder to detect.

Scientists are even observing the oscillations directly. At the KEK laboratory in Tsukuba, Japan, scientists have been creating a beam of muon neutrinos that they attempt to detect at the Super-K detector, about 250 kilometers away. The detections are very rare; they picked up about forty-four of those muon neutrinos in two years, though they should have seen about sixty-four. Again, it is a hint that neutrinos change flavors. Neutrinos have mass. But how much mass? Nobody is sure yet.

As physicists rapidly zero in on the properties of the neutrinos, as they figure out their mass and the way that they oscillate, they will finally understand the most elusive particles known to science. There are still many properties left to understand. For instance, some theories suggest that other flavors of neutrino exist that are unassociated with a tau particle, electron, or muon; these so-called sterile neutrinos are looking increasingly unlikely and will probably be ruled out. Also, some physicists believe that a neutrino has the bizarre property of being its own antiparticle. A particle with this property is called a Majorana neutrino; the classical particle is a Dirac neutrino. Majorana neutrinos should produce some exotic decays that scientists have not yet credibly detected. Most of these questions should be answered by 2010.[15]

Cosmologists nevertheless have already taken note: neutrinos oscillate, therefore they have mass. Since they have mass, they are able to exert a tiny influence on the curvature of spacetime; they must be accounted for in the fraction of matter in the universe, Ω_m, which scientists think is about 35 percent of Ω. However, neutrinos are not baryonic matter, like atoms, so they don't belong to the baryonic fraction, Ω_b, which accounts for only about 5 percent of Ω. Thus, neutri-

15. In 2001 the Super-K detector had a major accident. One of the light-detecting tubes shattered, and the shockwave wiped out most of the other tubes, blinding the Super-K detector. It will take several years for it to come back up to full speed—a heavy blow to neutrino physics. However, progress is so rapid, especially now that the Sudbury facility is operating, that scientists should still understand these mysteries by the end of the decade.

nos belong to the exotic matter, the nonbaryonic stuff that accounts for the rest of Ω_m. But do they solve the mystery of the
missing matter? Are they heavy enough to outweigh the rest
of the matter in the universe?

Surprisingly, the answer seems to be no. In 2002 the 2dF
survey released results that implied that neutrinos can make
up only a small fraction of the dark matter in the universe. It
is just a tiny slice of dark matter, but because dark matter is
so much more abundant than ordinary matter, the combined
mass of neutrinos is as much as all the stars and galaxies in
the universe—but not enough to solve the missing matter
problem.

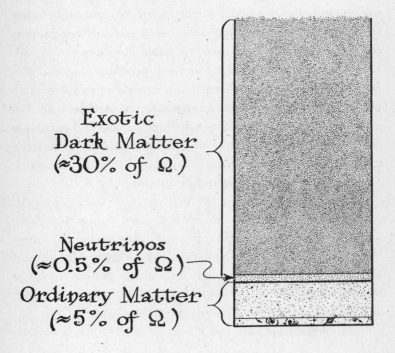

Exotic
Dark Matter
($\approx$30% of Ω)

Neutrinos
($\approx$0.5% of Ω)

Ordinary Matter
($\approx$5% of Ω)

Why, then, spend so much time talking about neutrinos?
Though it is true that there are not enough neutrinos to make
up all of the nonbaryonic portion of Ω_m, they seem to be
roughly as heavy as all the matter that astronomers can see.

These invisible particles are truly a concrete example of exotic dark matter, and they probably share much in common with the other exotic matter out there. And yes, there is other stuff, even more exotic than neutrinos, out there.

Scientists have ruled out all the quarks and all the leptons in the standard model, including the neutrinos, as the main source of exotic dark matter. Yet a large proportion of Ω_m is still unaccounted for. Could this be the signature of matter beyond the standard model, stuff that isn't accounted for in the theory? Most scientists think so, and like neutrinos, this superexotic matter can only interact through the weak force. So in a sense neutrinos are a link between the standard model and the stuff beyond our present theories. When scientists understand neutrinos, they will have their first real understanding of the mysterious exotic stuff that makes up most of the matter of the universe.

By studying neutrinos, scientists are truly beginning to probe the nature of nonbaryonic matter. They are a significant fraction of the exotic dark matter in the universe, but they are far from the entire answer. For the rest, scientists are looking—believe it or not—for a particle more exotic still.

Chapter 10
Supersymmetry

[FEARLESSLY FRAMING THE
LAWS OF MATTER]

... when there are other unavoidable obstructions, it will be permissible to make diminutions or additions in the symmetrical relations —with ingenuity and acuteness, however, so that the result may be not unlike the beauty which is due to true symmetry.

—Vitruvius, *The Ten Books on Architecture*

There is something more exotic than neutrinos, something more difficult to detect than the nearly massless particles—or so many physicists believe. If they are correct, then the standard model, which describes all the known properties of matter and its interactions, is about to change dramatically. Physicists have conjured a theory about matter that sounds as if it were straight out of a *Star Trek* plot: it proposes that every particle has an as-yet-undiscovered doppelgänger, a shadowy twin *superpartner* that has vastly different properties from the particles we know. Though this theory replaces the standard model and doubles the size of the particle zoo—and proposes that a whole slew of particles must yet be discovered—scientists have quickly grown very attached to this theory: supersymmetry. If supersymmetry is correct, these so-far-undetected partner particles are probably the source of exotic dark matter, the nonbaryonic stuff that makes up almost all of the mass in the cosmos. In that case, the mystery of matter will be completely solved, and cosmologists will understand the composition of every single bit of mass in the universe.

However, there is a second aspect to supersymmetry. It doesn't just explain the nature of matter. It also extends the standard model to a time when the universe was much hotter and denser, a time shortly after the big bang, before the age of the quark-gluon plasma, when the known laws of physics fail to apply. If supersymmetry is correct, scientists will be able to analyze the nearly immediate aftermath of the birth of the universe, giving them a glimpse of the first few fractions of a fraction of a second after the fiery cataclysm.

If these shadowy partners stay undetectable, then the theory of supersymmetry would be merely a mathematical toy. Like the Ptolemaic universe, it would appear to explain the workings of the cosmos, yet it would not reflect reality. The next decade will make or break supersymmetry. Scientists will confirm it, finding the first supersymmetric particle, or they will fail and give up the theory for good. A particle accelerator already in operation has a chance of flushing a supersymmetric particle out of hiding, and a second accelerator, already under construction, is almost guaranteed to find one—if supersymmetry is correct. In the next decade, scientists will confirm a revolutionary theory that can reveal the state of the universe at the time of the big bang and can reveal the hiding place of all the exotic matter in the cosmos, or they will go back to the drawing board.

Never fear. The last few chapters were a barrage of baryons, leptons, and mesons, but we are done with the details of particle physics. All the tools of the standard model are at our disposal. Considering how many particles we had to learn about to get a grasp of the standard model, it probably seems like an act of masochism to double the size of the model, but that is exactly what scientists are trying to do. The fundamental tenet of supersymmetry is that every particle in the standard model has a supersymmetric twin. (The supersymmetric electron is known as the selectron; supersymmetric quarks are squarks. There are sneutrinos, photinos, gluinos, winos, and zinos.) Each *sparticle* is related to its twin particle, but it is not the same. What's more, the mere existence of these supersymmetric particles subtly alters the predictions of the standard model.

Doubling the size of the standard model seems as though it should make the science of particle physics twice as complex. It doesn't. In fact, to physicists, supersymmetry looks almost the same as the standard model; all the different phenomena in the standard model (and in supersymmetry) are really different facets of a single mathematical object: a symmetry group.

Symmetries are a powerful tool for figuring out the underlying structure of an object. In a sense, a symmetry is simply a fancy term for a pattern, and science is, at its heart, a search for patterns. A crystal like a diamond is highly symmetrical because its atoms fall into a regular pattern, and because this pattern is so regular, it is easy for a scientist to describe. (And hard for a scientist to purchase, but that's another story.) The standard model of particle physics is also a description of symmetries; specifically, it is a mathematical object that incorporates all of the symmetries that govern the behavior of subatomic particles.

That is a pretty abstract statement. As a concrete analogy, imagine that you are handed a die. Though you don't know the shape of the die, you are asked to figure out its underlying structure: how many sides does it have? To determine this, the only thing you are allowed to do is to perform some experiments; roll the die a few times and see what happens. Af-

ter a little while, after a few rolls of the die, you discover that
the numbers 1, 2, 3, and 4 come up over and over again.
Maybe they don't turn up with the same frequency; the die
might be a little biased, but that can be ignored for the moment. By noticing that the numbers 1, 2, 3, and 4 appear and
reappear, you've spotted a pattern, and you can make a hypothesis. For example, you might tentatively assume that the
die has four sides, like a pyramid.

So far, so good; you've made a model of the underlying
structure of the die. This is what scientists do when they
make their own models. For instance, the many chemical elements once seemed like a jumble of somewhat-related com-
pounds with no underlying structure. But with the discovery
of the periodic table—the set of symmetries governing the relationships among the elements—chemists were able to fuse
those individual objects into a greater scheme, a single, symmetric object. By making the hypothesis that the die is shaped
like a pyramid, you are doing the same thing. You can treat
the outcomes 1, 2, 3, and 4 as different facets of the same
pyramid-shaped object. It is a subtle point, but by doing this,
you have shifted your perspective. Instead of trying to understand what causes each number to come up when you roll the
die, you can try to understand the shape of an underlying object that governs each of those four outcomes.

This is what particle physicists do when they try to understand the structure of the subatomic world. Of course,
their experiments aren't die rolls. (In fact, physicists are no-
toriously averse to rolling dice.)[1] Physicists observe the properties of subatomic particles by spotting their tracks in cloud
chambers, or they smash objects together and watch the re-

---

1. An apocryphal story about the American Physical Society tells how the society of
physicists held one of its annual meetings in Las Vegas. The physicists, who are ex-
perts at calculating odds, refused to play at the casinos. They knew that the casinos
had a statistical advantage and were guaranteed to win in the long run. As a result,
the story goes, the casinos got quite peeved at the loss of revenue from an entire convention of nongamblers, so physicists were declared personae non grata and have
not been allowed to hold another conference in Vegas.

sulting spray of particles. Just as you can get different numbers each time you roll a die, physicists get a different spray of particles each time they smash objects together. Some particles are common and easy to create, like the electron, and some are harder to make and rarer, like the muon.

After performing thousands and thousands of experiments, scientists figured out a mathematical structure that could explain their observations. They figured out the shape of their die, and this "shape" is the foundation of the standard model. The analogy is a bit deeper than you might expect. In a mathematical sense, the standard model really is a shape of sorts.

The first bits of this shape were discovered in the 1960s. Murray Gell-Mann proposed the theory of quarks. His essential achievement was that he sensed an underlying shape to the menagerie of particles. When he looked at the eight baryons that (we now know) consist of up, down, and strange quarks, he saw a pattern that he dubbed the *eightfold way* after the Buddhist tenets for leading a good life and achieving enlightenment.[2] The eightfold way, this pattern Gell-Mann saw, is known as a symmetry group. The standard model is a grander scheme, a larger shape that incorporates Gell-Mann's quarks and all of the other known fundamental particles in the universe.

All the known particles in the universe sit on the abstract shape of the standard model. This shape cannot be visualized so easily because it is a seven-dimensional object,[3] but the

2. Unfortunately, some popularizers of physics have played up the connection between Eastern philosophy and particle physics to an absurd extreme.

3. Mathematicians don't usually think of these objects as shapes, even though the concept of a *group* is closely tied to the symmetries of an object like a pyramid or a cube. Most groups, like the standard model's—which formally has the structure known as $SU(3) \times SU(2) \times U(1)$, the $SU(3)$ part being the portion of the group that has to do with quantum chromodynamics—are too complex to be described by the symmetries of a three-dimensional object; instead, the mathematician has to go to higher dimensions. This has nothing to do with whether the particles, or the universe itself, have more than three (or four) dimensions. It just has to do with the abstract object that is associated with a group. The dimensions are a mathematical formalism and nothing more.

symmetries inherent in that shape define the rules of nature that define the subatomic world.

Going back to the die-rolling analogy, the more you roll the die, the more confident you become that your hypothesis about the shape of the die is correct. However, something un-expected might happen and mess up your understanding of the die. For instance, imagine that, after rolling the die lots and lots of times and seeing the numbers 1, 2, 3, and 4 crop up over and over again, you are quite convinced that you are rolling a four-sided die, a pyramid whose four triangular sides are inscribed with the numbers one through four. Then, casu-ally, without expecting anything out of the ordinary, you throw the die very hard and up pops the number 5. This sur-prising outcome blows away your pyramid-shaped model of the die. There is no way a four-sided die can have a fifth out-come. You have to refine your model; you must extend it to explain the new result. Perhaps the die has six sides instead of four. If a six-sided model is correct, and you have only seen five sides, simple arithmetic tells you that a number is left over, waiting to be seen.

This is precisely what happened in particle physics. When Gell-Mann proposed his quark theory, scientists had not seen all the sides of his die, so to speak. One side of Gell-Mann's abstract shape had not been observed; one particle was missing. There was no experimental evidence that the missing particle existed, but it *had* to be there if Gell-Mann's theory was right. The symmetries of Gell-Mann's abstract, symmetric, object demanded it. In 1964, physicists at Brook-haven National Laboratory found the missing particle, the omega minus. Gell-Mann was right. His shape had predicted the existence of a new particle.

The standard model is the current state of the art. It con-tains Gell-Mann's object and more, and it describes the inter-actions in all the experiments done over the past forty years with incredible accuracy. The standard model encodes all the scientific knowledge about the enormous die that governs the interactions of matter. This die is not evenly weighted; some of its "faces" are harder to see than others. However, the more times scientists roll the die in their experiments, and the more energetically they roll it (to overcome the uneven weighting), the more confident scientists are that the standard model is essentially correct. After billions of die rolls, scientists are (on the whole) quite happy with the standard model. However, there are a few loose ends.

One of the problems is that of *unification.* At sufficiently high energies, the electromagnetic force and the weak force become different aspects of the same force: the electroweak force. Sand and glass are actually the same substance, but this only becomes apparent if you heat them enough. Similarly, the electromagnetic and weak forces are two aspects of the same thing but only behave that way at very high energies. Unfortunately, when you consult the standard model, the strong force doesn't "unify" quite as easily as the other two forces. Scientists, naturally, hope that the strong force and eventually gravity wind up being different aspects of the same underlying phenomenon, making them easier to understand. The standard model does not account for the unification of the strong and electroweak forces, but supersymmetry, the extension of the standard model, does.

That is just one reason for looking at supersymmetry. There are others. For instance, the standard model is unable to account for the mass of particles, but supersymmetry naturally accounts for particle masses; the Higgs boson springs to life, full grown, out of supersymmetry's equations. What's more, supersymmetry also naturally solves some of the outstanding puzzles in cosmology, such as identifying the exotic dark matter and determining the forces that drove the expansion of the early universe. A cosmos governed by the rules of supersymmetry has a lot fewer unanswered questions than one governed by the standard model. However, supersymmetry has an obvious disadvantage because it doubles the size of the particle zoo.

The standard model works extremely well, so it should be no surprise that supersymmetry extends it, rather than replaces it, just as the standard model extended Gell-Mann's eightfold way. This means that the "shape" of the standard model has to be included in the larger "shape" of supersymmetry. Unfortunately, the mathematical rules that govern the symmetries of shapes, the theorems of group theory, require that the new, extended shape must be at least twice the size of the standard model's. Since this abstract shape is merely a representation of the particles that govern the subatomic world, doubling the size of the shape means doubling the number of particles. So with all the advantages of supersymmetry, physicists get the disadvantage of a huge collection of undetected particles.

However, that disadvantage can instantly become an advantage if a new particle is discovered. It has happened in the past: Dirac's postulate of an antimatter electron became an asset rather than a liability as soon as Carl Anderson found the wispy trace of an antielectron in his cloud chamber. So scientists reserve judgment. In the meantime, they are on the lookout for hints of a supersymmetric particle; if it is found, it will imply that supersymmetry is correct and will double the size of the particle zoo. So far, physicists have had no solid success, in part because the effects of supersymmetry can be very small.

When you postulate that a die has six sides rather than four, you are altering your predictions for the probability of seeing a particular number on a given roll. When there is a chance of rolling a 5 or a 6, no longer are you 100 percent guaranteed to roll a 1, 2, 3, or 4. In the same way, supersymmetry changes the probability of "rolling" any particular particle. Since the properties of the individual particles are intimately related to these "probabilities," changing the probabilities, as supersymmetry does, also slightly alters those properties. If scientists spot a property of a particle that is not quite what the standard model predicts, it might be the signature of supersymmetry's subtle influence.

In early 2001, scientists at Brookhaven were extremely excited about a possible hint of supersymmetry. They were measuring a property of the muon known as its *magnetic moment,* which describes how strongly a particle twists in a magnetic field, when they realized that something was amiss. For three years, physicists had fed muons into the field of a fourteen-meter-wide superconducting magnet and forced the particles to twist around in a circle. When the scientists analyzed how much the muons twisted in that field, they found that their value disagreed with the standard model's prediction by about four parts in a million. It might not seem like much of a disagreement, but the standard model is so successful that even such a tiny variance might indicate the model's failure — and a hint of supersymmetry. If the experiment was correct, then the discrepancy might be the signal of an unseen supersymmetric particle influencing the properties of the muon. The discovery excited physicists all over the world. However, the excitement was short-lived. By December, physicists found that the standard model's prediction was done incorrectly; the two physicists who did the calculation introduced a superfluous minus sign into their calculations, messing up their value for the standard model's prediction. When this tiny error was corrected, the discrepancy between experiment and theory dropped dramatically, though additional data made the difference greater yet again. As of the end of 2002, it seemed that a reasonable anomaly remained, but it was too small for the Brookhaven physicists to claim that they have seen a sign of supersymmetry.

This is not the first frustration for scientists searching for supersymmetry. Some sightings have disappeared, and others are in limbo. For instance, physicists working on four experiments at CERN spotted a tantalizing signal that might have been a trace of a supersymmetric particle. The experiments used the Large Electron-Positron collider (LEP), an enormous accelerator that smashes electrons and antielectrons together while detectors measure the spray of particles that fly off in the enormous bursts of energy that result. The standard model predicts that particles will be created in certain proportions; there will be so many up quarks, electrons, neutrinos, and so forth. The LEP experiments were concerned with the number of tau particles. They found that at low energies—when the die wasn't thrown very hard—the number of taus matched the predictions of the standard model. However, at higher energies, they saw more tau particles than the standard model predicted: 228 tau reactions of a certain type where 170 were expected. Again, it looked like a possible signature of supersymmetry, but scientists weren't able to make a very firm statement. The only way to determine whether the excess taus were a true sign of supersymmetry or merely a statistical fluke was to acquire more data with the LEP—roll the die more times—and see whether the discrepancy got larger or smaller. Unfortunately, the physicists released this result a few months before the LEP was torn up in 2000 to make way for a new machine. The LEP experiment was dead. But it was killed for a good reason. The accelerator was destroyed to make way for a machine that will put the supersymmetry question to rest once and for all.

The tunnels already carved out for the LEP are being refitted with even more-sophisticated magnets. They will be the heart of the Large Hadron Collider (LHC), the next-generation particle collider.

If you stand in the caverns of the LHC, which circle under peaceful meadows in the Geneva suburbs, you are dwarfed by the immensity of the detectors. Merely getting the machines down in the tunnels was a tremendous task. At one site, the builders had to tunnel through an underground river to lower one of the detectors into place. So they froze the river with liquid nitrogen, burrowed through the ice, and let it flow again once they had built an access shaft.[4] As you can imag-

4. The accelerator and the neutrino detectors are underground for different reasons. Neutrino detectors must be shielded from cosmic rays; people must be shielded from particle accelerators, which shed a great deal of radiation as the particles swing around in circles. (This is a phenomenon known as synchrotron radiation.)

ine, such a project costs billions of dollars, and it is already significantly overbudget. Nevertheless, the LHC is a magnificent machine, and it will be the centerpiece for high-energy physics in the century's second decade.

The LHC will be so powerful that it will have the energy to discover a supersymmetric particle; if it doesn't, then the theory of supersymmetry will most likely be ruled out. The LHC is scheduled to come online in 2007, and after it runs for a few years, we will know, once and for all, whether supersymmetry is a reflection of the way the world works, or whether it is just a pipe dream of theorists looking for an aesthetic, unified universe.[5]

If we're lucky, we won't have to wait that long. In 2002, researchers at the Fermi National Accelerator Laboratory (Fermilab) in Batavia, Illinois, were trying to work the kinks out of another powerful accelerator—the Tevatron—in hopes of beating the LHC to the punch. Though the Tevatron accelerator was having some serious teething problems after a $260 million refit, scientists believe that the machine, which smashes protons and antiprotons together at high energies, may already have the power to find a smoking gun for supersymmetry.

The confirmation of supersymmetry not only will double the size of the particle zoo and give physicists a whole new realm to explore, but will also give cosmologists a new suspect in the hunt for exotic dark matter. Most supersymmetric particles are expected to be unstable, decaying in less than a second to other, more stable forms of matter. Yet at least one of these sparticles must be relatively stable—the lightest supersymmetric partner, LSP. Scientists do not know what the LSP is; it may well be a mixture of sparticles, just as neutrinos are blended together, constantly oscillating and changing flavors. However, more than twenty years ago, the scientists who studied supersymmetry realized that if the theory is correct, a

5. Thanks to the cost overruns, 2007 is looking optimistic by a year or two.

lot of LSPs had to be floating about in space. In fact, the mass tied up in LSPs should outweigh the amount of baryonic matter in the universe. For a while, this sounded like a ridiculous proposition, but as the previous chapters have shown, cosmologists have been forced to accept the existence of a whole lot of exotic dark matter. Supersymmetry would be a beautiful and succinct explanation for exotic dark matter, so many scientists naturally think that LSPs are the source of exotic dark matter. In addition, LSPs are better than neutrinos for explaining the bulk of dark matter, because a hypothetical LSP's properties account for the structure of the universe better than those of the neutrinos. This is because neutrinos are "hot" dark matter, while LSPs are "cold."

Even though neutrinos have mass (which can be converted to energy, thanks to Einstein's famous $E = mc^2$), they have very little of it. Most of their mass-energy comes from their incredible speeds: neutrinos born in the early universe would tend to travel at close to the speed of light. These fast-moving particles are considered "hot," like the fast-moving molecules in a tub of boiling water. On the other hand, LSPs would be much more massive than neutrinos — tens to hundreds of times heavier than the proton — so more of their energy would be tied up in their mass rather than in their motion. These slow-moving particles would be "cold," like the relatively lethargic molecules in a sheet of ice.

Cold dark matter and hot dark matter have different influences on the development of structures in the universe. In theory, hot dark matter tends to make larger structures, like galaxy superclusters, form earlier than smaller structures, like stars and galaxies. Cold dark matter tends to have the opposite effect, causing the smaller structures to form before larger ones, and this seems to be the way things in the universe formed, with stars building into galaxies building into galaxy clusters.

Scientists are uncertain about how much of the exotic dark matter is cold and how much is hot, but the argument is leaning toward a preponderance of cold matter rather than hot matter. If this is the case, then neutrinos' contribution to exotic dark matter has to be exceeded by the contribution of some nonneutrino particle, and as we saw, the neutrinos are the only particles in the standard model that can make up a substantial fraction of the exotic dark matter in the universe. If the cold dark matter theories are correct, then there has to be physics beyond the standard model; there has to be another, undiscovered particle that is responsible for cold dark matter.

Supersymmetry is the best hope, and physicists are searching desperately for the LSP, the lightest supersymmetric partner, which might hold the key to the missing exotic matter in the universe. If the bulk of dark matter is not made up of the LSP, then scientists are, once again, stuck with a die with too few faces. On the other hand, if the LSP is responsible for cold dark matter, then physicists, in unraveling the mysteries of matter, will have solved, at least on the level of particle physics, the problem of the missing stuff in the universe. Supersymmetry will triumph, and the components of Ω_m would be understood. But there is one remaining problem for cosmologists. They have yet to spot a chunk of dark matter, and until they do, they can scarcely celebrate victory.

But the time of that victory may be at hand. Astronomers are learning to see the invisible. When they do, they will have solved the final mystery of matter.

Chapter 11
Seeing the Invisible

Oh leave the Wise our measures to collate

One thing at least is certain, light has weight

One thing is certain and the rest debate

Light rays, when near the Sun, do not go straight.

—ARTHUR EDDINGTON, 1919[1]

Whether matter is governed by supersymmetry or some other extension of the standard model, a complete theory that describes the subatomic world must identify the particles that constitute the missing dark matter. Cosmologists have concluded that some of this matter is dark and baryonic, and some is made of neutrinos; the remainder—most of the missing matter in the universe—is neither. Maybe it is an exotic supersymmetric particle, maybe it is something else, but when scientists figure out the correct theory of matter, they

1. Though light has momentum, it does not actually have mass, so the poem is a bit misleading. Such is the nature of poetic license.

will only know the sort of particle that makes up the missing substance. The correct theory of matter will not reveal where the missing matter is hiding. Astronomers and cosmologists, not particle physicists, are the ones who are going to find out where it resides.

Of course, dark matter is dark—dark as space itself, and invisible to telescopes—but it must exist. The structure of galaxy clusters, the motions and distributions of galaxies, and the fine details of the cosmic background radiation all imply that the universe is filled with dark matter. However, as long as scientists are unable to see it directly, they cannot truly claim to understand it.

The veil is finally beginning to lift. In the past few years, astronomers and physicists have been sniffing out the hiding places where dark matter, both baryonic and exotic, might be. Armed with orbiting telescopes, underground laboratories, and a variety of other instruments, they are finally finding the signatures of the invisible mass. The first sightings are already beginning to dribble in. By detecting the subtle warping of space and time, physicists are able to pinpoint small, dark objects floating about in the invisible halo of matter about our galaxy. On a grander scale, this warping is helping physicists detect the vast collection of invisible matter that holds galaxy clusters together. X-rays that stream from galaxies and clusters are revealing the nature of the dark matter particles. A forest of absorption lines in light striking telescopes is giving away the presence of clouds of gas in deep space, far from the illumination of galactic light. If scientists are lucky, they might even be able to capture some of the exotic dark matter in specially designed traps buried under the earth. Though it is invisible, dark matter will soon divulge its secrets.

Even as scientists close in on dark matter, they are also learning about the "dark age" of the universe: the time between recombination, when light was freed from its cage of matter, and the era of *reionization*, when the first stars ignited and galaxies began to blaze forth. Astronomers are beginning to pierce remotest darkness. Soon, matter will have nowhere left to hide.

Since there are two types of dark matter, the search for it has two branches. The first is the search for baryonic dark matter, the nonluminous stuff made of quarks. Though physicists know what most of this "ordinary" matter is made of, it remains undetected, so they must sniff out its hiding places. The second branch is the search for exotic dark matter, the stuff not made of quarks and which outweighs baryonic matter by about six to one. Each branch of the search uses different instruments and techniques.

It's not easy to spot something that is by definition invisible, but it's not impossible. For instance, scientists have an ever growing catalog of massive collapsed stars known as black holes, even though a black hole is so massive that even light cannot escape its gravitational pull; it absorbs light rather than emitting it.[2] This means that the black hole itself is invisible; any radiation that comes too close to the point of no return, the event horizon, disappears down the maw of the collapsed star. How can scientists spot something that swallows light?

Even though a black hole is practically invisible, astronomers can infer its presence from the effects it has on spacetime itself. It is a much trickier observation to make than simply looking for a glowing object in the sky, but it can be done. For example, Andrea Ghez, an astronomer at UCLA, uses radio telescopes to study the motions of stars near the center of our galaxy. By watching how those stars move, she is really measuring the curvature of spacetime — the strength of gravity — in the heart of the Milky Way. When she calculated the curvature of spacetime from those stellar

2. Technically, all black holes emit something known as Hawking radiation. Hawking radiation is too feeble to detect directly. There is another way black holes can "emit" light, which is described in the next footnote.

motions, Ghez realized that the stars are wheeling about an invisible, supermassive object that weighs more than two and a half million times as much as our sun. The black hole at the center of the Milky Way, dubbed Sagittarius A° (the Milky Way's core is in the constellation Sagittarius), cannot be seen directly, but Ghez was able to find it because of the effect it has on spacetime, on the stars orbiting it.[3]

Ghez's technique is quite similar to what Vera Rubin did when she made the first compelling case for dark matter. Rubin's measurements of how the stars spun around the center of the Andromeda galaxy were really a measurement of the curvature of spacetime—an indirect measurement, but a measurement nonetheless. The stars' motions allowed her to "weigh" the galaxy and find out where its matter resides. When the calculated weight didn't match the weight inferred from visible stars, Rubin realized that there had to be an invisible halo of matter, a halo that increased the curvature of the spacetime sheet. This technique was old even when Rubin used it decades ago. As early as 1933, astronomer Fritz Zwicky used the same technique and noticed a few discrepancies in the motions of galaxies in a galaxy cluster; though his evidence was not terribly strong, Zwicky had caught the first whiff of dark matter. So, analyzing the motions of objects to measure the curvature of spacetime is nothing new.

However, there *is* a new technique for measuring the curvature of spacetime; astronomers can now see that curvature almost directly. Telescopes have only just become sensitive enough to use this technique, gravitational lensing, on a large scale. Astronomers can now measure how gravity bends the light of distant objects and calculate the gravitational pull of invisible matter. Gravitational lensing has already begun to reveal the dark matter surrounding our own galaxy, and in

---

3. In fact, supermassive black holes like Sgr A° aren't always invisible, because they are messy eaters. Even as they swallow matter and energy, they spew some out in bright jets that are sometimes visible half a universe away. Sgr A°, as it happens, is oddly quiet.

2000, scientists announced that they had used gravitational lenses to spot the first dark matter objects in a distant galaxy. With this technique, scientists will soon figure out where dark matter resides in the halo about the Milky Way. Better yet, they will get a much better handle on what makes up the baryonic fraction of the matter in the universe, Ω_b.

Though scientists have only recently been able to take advantage of this technique, the story of gravitational lensing is more than eighty years old. In fact, the first gravitational lens was what made Einstein a celebrity. In 1919, Arthur Eddington journeyed to the island of Príncipe, off the coast of Africa, to test one of Einstein's predictions: that the warping of spacetime by a massive object, like the sun, bends light as a lens does. According to the general theory of relativity, a star's light that passes close to the sun, deep into the dimple that the sun makes in the rubber sheet, should be warped by the curvature of spacetime. As a result, the star should appear in the wrong place in the heavens. Its apparent position in the sky should be somewhat altered by the gravitational pull of the sun. Since the sun is so bright, it was impossible to spot stars so close to the flaming disk—except during a solar eclipse. That is precisely what Eddington was after: in the early afternoon of March 29, 1919, the shadow of the moon traversed the island of Príncipe. A total solar eclipse blotted out the sun for a few minutes, allowing Eddington's team to measure the positions of stars close to the sun. After heavy rains, the skies let up just enough for Eddington to snap some photographs:

> The rain stopped about noon and about 1.30 . . . we began to get a glimpse of the sun. We had to carry out our photographs in faith. I did not see the eclipse, being too busy changing plates, except for one glance to make sure that it had begun and another half-way through to see how much cloud there was. We took sixteen photographs. They are all good of the sun, showing a very remarkable prominence; but the cloud has interfered with the star images. The last

few photographs show a few images which I hope will give us what we need.[4]

If Einstein's theory was right, the stars near the sun would appear in slightly different places from where they normally appear. This is, of course, precisely what Eddington saw. The stars were not in their proper places. Eddington and his team had seen the curvature of spacetime almost directly. They saw the warping of light around a heavy object: a gravitational lens.

More than seven decades later, one type of gravitational lensing, known as microlensing, began to reveal the distribution of dark matter in our own galaxy. Just as Eddington's gravitational lens was caused by the sun—a relatively small, compact chunk of matter, at least on a galactic scale—microlenses are subtle warpings of light caused by small, dense clumps of matter. (Other types of gravitational lenses, such as those caused by entire galaxies, are considerably larger and more diffuse.) These microlenses are the signature of dark matter in the halo that surrounds our galaxy. Scientists have spotted the tremors of warped light and found dark matter surrounding us.

These hunks of matter are known as MACHOs: massive, compact halo objects. Nobody is quite sure what these things are; they could be burned-out stars or brown dwarfs, stars too light to ignite their fusion engines. Whatever they are, they have mass. Anything with mass warps the fabric of spacetime and affects the passage of beams of light from background objects. Conveniently, there are lots of background objects that can reveal the presence of a MACHO. The best are stars in a small, nearby galaxy known as the Large Magellanic Cloud. (The Milky Way has a couple of small "satellite" galaxies that orbit just outside the Milky Way, trapped by its gravitational pull. The Large Magellanic Cloud is the biggest of these.) Visible from the Southern Hemisphere, the Large Magellanic Cloud lies just outside our galaxy and beyond the

4. Quoted in The MacTutor History of Mathematics Archive.

dark matter halo that holds the Milky Way together. It is also a tool that scientists use to find dark matter.

As a MACHO passes in front of one of those stars, the gravitational pull of the MACHO twists the light so that more of it is focused on the Earth. As Earth receives more and more light from the star, the star appears to brighten, and as the MACHO moves away over the course of a few weeks, the star dims, once again, to its original luminosity. So the brightening and dimming of a background star is a pretty good signal that a MACHO has passed between the background star and our solar system.

An international team of astronomers and astrophysicists use telescopes in Australia and the United States to find these microlenses—to spot MACHOs. This collaboration, dubbed the MACHO Project, keeps an eye on stars in the Large Magellanic Cloud and other background stars, measuring their brightnesses over and over and over again. (Several other groups also do the same thing, such as the aptly named OGLE collaboration.) The MACHO Project began in 1993, and since then it has spotted hundreds of these microlensing events; nowadays, astronomers spot one almost weekly. The flood of data is allowing astronomers to chart where the dark matter resides in our galaxy—and may allow them to figure out what it is made of; black holes distort light in a slightly different manner from the way brown dwarfs distort it, so astronomers might be able to tell what sort of massive bodies are floating about in the halo. In addition, microlenses may even reveal the structure of distant galaxies.

In early 2000, two Dutch astronomers spotted the signs of microlensing in a distant spiral galaxy, microlensing that revealed the presence of dark matter. The galaxy is very much like our own, and astronomers see it edge-on. Behind it, a quasar shines brightly.[5] The light from the quasar is bent

5. A quasar is a "quasi-stellar object," a very distant, small, bright source of energy. Scientists now think that a quasar is a galaxy with a massive black hole in the center that spews out enormous amounts of light.

by the objects in the galaxy, which act as gravitational lenses. But the spiral galaxy does not bend that light uniformly, and this caught the astronomers' attention. The light that passes through the visible part of the galaxy is fairly steady; it doesn't flicker much. However, the light that passes in the region where the galaxy's dark matter halo should be is flickery and unsteady. This flickering may be the sign of MACHOs in the distant galaxy's dark matter halo; even though the microlensing events are too faint to be seen individually, the collective effect might make the light from the distant quasar twinkle.

It is still too early to tell precisely what microlenses and MACHOs will reveal about the nature of baryonic dark matter and its distribution in galaxies, but it seems that microlenses might hold the key to the distribution of baryonic dark matter in galaxies.

Microlensing is not the end of the story, though. Microlensing reveals the presence of baryonic dark matter in large clumps, but it won't reveal pockets of diffuse gas or of exotic dark matter. A microlens is caused by a compact, dense object (like a MACHO) that creates a nice, small, highly curved lens that is visible from relatively nearby. On the other hand, a large, diffuse object like a gas cloud or a galaxy makes a broader, less curved lens that is only visible from much farther away.

A gravitational lens can have different strengths and create different effects, just as ordinary lenses do. Microlenses, such as those produced by MACHOs in the galactic halo, merely make stars appear to brighten and dim. But larger lenses can create much more dramatic effects. If an enormous amount of matter and a distant bright object are positioned in the right way, astronomers can see double. Thanks to the bending effect of the matter in the lens, light from the background object can take two or more curved paths to Earth. For each of those distinct paths, observers see a blotch in the sky—multiple images of a single object. Astronomers are using these broader lenses to weigh not only galaxies, but clusters of galaxies, the most massive things in the universe.

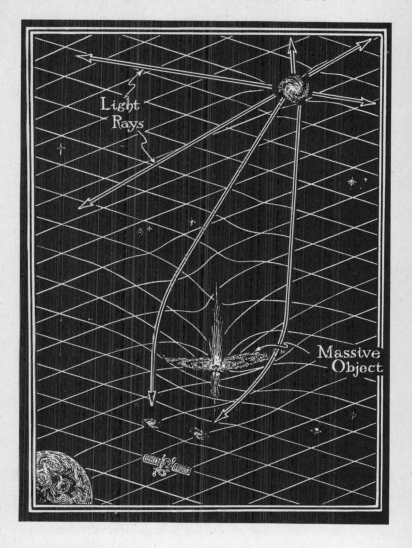

Gravitational lensing

In 1979, astronomer Dennis Walsh and his colleagues spotted the first of these enormous lenses; thanks to the enormous gravitational pull of a massive object he saw two images of the same bright quasar in the sky. But gravitational lenses aren't always so obvious. If the curvature is not too dramatic or the background object is not positioned in just the right place, a lens will not produce two images of the background object. Instead of making multiple images like a "strong" lens, a "weak" lens merely distorts the appearance of background objects, smearing them out into arcs of circles. Both strong and weak gravitational lenses are revealing where dark matter resides—both the baryonic and exotic varieties.

Astronomers can use a gravitational lens to weigh a massive object like a galaxy or a galaxy cluster. The more dramatic the distortion—the stronger the lens is—the more matter there is and the more tightly packed it is. If the background objects are good enough to illuminate the lens's structure, researchers can make a very detailed map of where the matter in a galaxy or galaxy cluster is, including its dark matter. In mid-2001, for example, scientists at Bell Laboratories discovered a previously unknown cluster of galaxies, one that was invisible to telescopes. Though it could not be seen directly, the dark cluster was spotted 3.5 billion light-years away because it bent the light from even more distant objects.

The curvature of spacetime depends on *all* the mass in the cluster, so the lens reveals the location of exotic dark matter in a galaxy cluster as well as the baryonic stuff. Because of this, gravitational lenses are becoming a powerful tool for cosmologists who are trying to pin down the distribution of dark matter in the universe. Unfortunately, it is relatively hard to spot gravitational lenses, because a background object must be positioned in just the right place for an astronomer to see the subtle warping of light. Luckily, scientists have other techniques to look at dark matter, and these techniques don't even need a star or a quasar in the background.

One of the most promising techniques for figuring out the

contents of galaxy clusters uses the cosmic background radiation as illumination rather than bright objects in the background. Measurements of galaxy clusters using the cosmic background radiation are a little trickier than those using discrete light sources and rely on different scientific principles. One of the techniques depends upon the *Sunyaev-Zel'dovich effect,* named after the Russian physicists Rashid Sunyaev and Yakov Zel'dovich. Sunyaev and Zel'dovich realized that if a galaxy cluster has enough hot matter, it has a lot of very fast electrons. When a photon bumps into one of these electrons, the electron gives the photon an extra kick of energy. Depending on the circumstances, this distorts the spectrum of the cosmic background radiation, creating a hot spot that wouldn't ordinarily be there.

Already, dozens of scientists are trying to use the Sunyaev-Zel'dovich effect to measure the size and distance of clusters, this providing another method for estimating the rate of Hubble expansion of the universe. These measurements are only becoming possible now that the new generation of microwave telescopes — Boomerang, DASI, and the MAP satellite — is gathering data precise enough to actually see the Sunyaev-Zel'dovich effect. The new microwave telescopes (along with their enormous radio-detecting brothers) have opened up a whole new realm for astronomers to explore and have made the Sunyaev-Zel'dovich effect a practical tool for determining the structure of distant objects.[6]

X-ray astronomers too are using a new telescope to figure out the nature of dark matter. The Chandra X-ray Observatory, the newest x-ray telescope, was launched at the end of 1999 from the space shuttle *Columbia.* Chandra, the Hubble telescope for x-ray astronomers, now loops around the Earth in a highly elliptical orbit, high above the planet's x-ray—

6. In 2002, the National Science Foundation approved a $17 million South Pole telescope that will use the Sunyaev-Zel'dovich effect to find galaxies that are too dim or distant to be spotted by other means — and to map the distribution of matter in the universe.

absorbing atmosphere. It has already made its mark on cosmology.

John Arabadjis, an astronomer at the Massachusetts Institute of Technology, is using Chandra data to figure out the properties of dark matter in galaxy clusters, including exotic, nonbaryonic dark matter. In 2001 he published a study of a galaxy cluster known as EMSS 1358+6245, an object that was first discovered in 1968 by Fritz Zwicky (the astronomer who, thirty-five years earlier, published the first hints of the existence of dark matter). Arabadjis and his colleagues wanted to get more information about the matter inside the cluster, so they pointed the Chandra telescope at it for a little more than half a day. Since the hotter something is, the more x-rays it emits and the higher their energy, Arabadjis and colleagues were able to figure out the temperatures of various regions of the cluster by using Chandra's observations of the amounts and kinds of x-rays coming from those regions. These temperatures, in turn, revealed the amount of mass in those regions, because the temperature, the pressure, and the density of the mass are all interrelated.

What is particularly interesting about this type of observation is that it is revealing the nature of exotic dark matter as well as its distribution. Both strong and weak gravitational lensing provide a snapshot of a galaxy's matter distribution, but unlike microlensing, which can tell cosmologists something about the nature of MACHOs, they reveal little about the nature of dark matter beyond how it is distributed within the galaxy. Scientists get almost no information about the composition of this exotic stuff. The x-ray technique, though, can do what gravitational lensing can't.

From the temperature, pressure, and density of EMSS 1358+6245, Arabadjis and his team made a detailed picture of mass distribution, and from this mass distribution they were able to figure out some of the properties of the particles of exotic dark matter. For example, one theory about the particles said that bits of dark matter were "big," that they bumped into one another a lot, pushing one another quite a bit. This, in turn, made the exotic dark matter spread out. (It's kind of like urban sprawl; if your neighbors bump into you and bother you a lot, you're more likely to move into the suburbs to get away from them. If, on the other hand, your neighbors don't bother you so much, you can endure the crowds of the city.) This dark-matter-bumps-into-itself-a-lot theory is called *self-interacting dark matter;* and for various reasons, many astronomers liked it.[7] But Arabadjis showed that in EMSS 1358+6245, at least, the dark matter does not spread out as much as self-interacting dark matter should. In fact, Arabadjis was able to calculate how much a dark matter particle "annoys" its neighbors, something that physicists call the *cross-section* of the dark matter particles. It is something like the dark matter particle's effective size, and Arabadjis calculated that five grams of dark matter particles would take up no more space than a penny. It's a pretty crude number by particle physics terms, but it's enough to kill off the theory of self-interacting dark matter. More important, it shows that a cluster of galaxies many, many light-years away can reveal the properties of subatomic particles. These kinds of measurements will get better in the future, and physicists have even more tricks up their sleeves to figure out the nature of exotic dark matter. They even hope to trap it directly.

Perhaps the exotic dark matter hunters wanted to distinguish themselves from those who look for MACHOs, or maybe it was just a coincidence. In any case, the scientists who are trying to trap a piece of exotic dark matter have dubbed their quarry a WIMP, a weakly interacting massive particle. It is "weakly interacting" because it is thought to be affected by the weak force, though not the strong or electromagnetic forces; it is massive because it affects the curvature of the universe, making up the bulk of Ω_m. What is a WIMP? Nobody's quite sure. The LSP is a prime candidate. Already

---

7. It explained, for example, the rotation rates of dwarf galaxies very nicely.

some groups have claimed to have trapped WIMPs, but their claim is shady at best. Nonetheless, observatories are springing up around the world, even underneath the ice sheet of Antarctica. They hope to spot their first WIMPs by the end of the decade.[8]

A WIMP hunt is very much like a neutrino hunt, because WIMPs have so much in common with neutrinos. Neither a WIMP nor a neutrino feels the effect of matter all that much; indifferent to the strong and electromagnetic forces, both particles can travel through acres of matter without feeling much the worse for wear. But once in a great while, a neutrino or a WIMP will interact with a detector via the weak force, creating a telltale flash of light in its wake. Indeed, it is likely that an observatory designed to detect neutrinos will catch the scent of a WIMP as well. The trick is to be able to tell which of the many flashes in the detector are the signatures of WIMPs, which are from neutrinos, and which are from other phenomena.

Lots of things can confuse a neutrino detector. For instance, cosmic rays, highly energetic particles that come from beyond our galaxy, can hit a neutrino detector, causing a flash of light that can be mistaken for a WIMP or neutrino. As noted in chapter 9, neutrino detectors are placed deep under the earth so that almost every incoming particle will hit an atom of rock and get scattered or absorbed. Cosmic rays and even high-energy gamma rays, which can penetrate yards of concrete, stand no chance of going through tons and tons of rock. On the other hand, neutrinos and WIMPs can penetrate any massive barrier with ease; they barely interact with matter at all, so they will zoom through the barrier without any problem. A mountain of shielding will block everything except neutrinos and WIMPs.

8. There are a few other dark matter candidates other than WIMPs, such as an exotic beyond-standard-model particle called an axion, but WIMPs are clearly the favorite of the physics community.

That's only half the battle, though. Scientists must still tell the difference between the flashes caused by neutrinos and those caused by WIMPS. It is not an easy task, but it is possible. For example, researchers are looking for how the number of flashes in the detectors change with the seasons — the signature of the Earth's motion through a WIMP "wind."

Presumably, there are lots of WIMPs in the galactic halo that surrounds the Milky Way (the MACHOs don't seem to mind the company of their WIMPy neighbors). Since the solar system, in its long orbit around the center of the galaxy, moves through the halo, the Earth is constantly being buffeted by a wind of WIMPs. When the Earth, in its annual course around the sun, moves upwind (in June), it should be hit with more WIMPs than when it is traveling downwind in December. So, if physicists are spotting WIMPs in their detectors, they should see the number of flashes in their instruments wax and wane with the seasons.

This is precisely what physicists at the University of Rome claim to have seen at the Gran Sasso underground laboratory in Italy. Led by physicist Pierluigi Belli, the team said in 2001 that over the course of four years it had seen an annual increase and decrease in the number of flashes in their detector, a detector that is shielded by a kilometer and a half of mountain overhead. Other physicists are very skeptical of the claims, because other dark matter searches have failed to reproduce the results. It is a controversy that has been festering in the physics community. In June 2002 the EDELWEISS team, which uses a similar detector (buried under the French Alps rather than the Italian Alps), all but ruled out the Italian dark matter sighting; their detector, even more sensitive than the Italian one, should have spotted dark matter candidates if the Italians were correct. They didn't, all but killing the Italian claim.

The hunt for exotic dark matter is a high-stakes game; the first team to spot it will probably get the Nobel Prize. Several detectors just coming online, including a revamped version of

EDELWEISS, are hot on the trail, hoping to spot the evanescent signal of a WIMP. EDELWEISS will face lots of com-
petition from detectors all over the world.

One of these, AMANDA, is in the Antarctic wasteland,
where scientists are using the Antarctic ice as an enormous
neutrino detector. The Antarctic is a forbidding environment,
but it has one big advantage for WIMP and neutrino hunters:
a sheet of ice that is kilometers thick. This sheet of ice is a
huge chunk of mass, just as a mountain is, so a detector buried
underneath enough ice is also shielded from stray cosmic ray
particles. And instead of having to bore through a mountain
of rock, the AMANDA members simply had to melt their
way down, though in the Antarctic, merely getting enough
fuel to melt the ice was an enormous task. Nevertheless, the
AMANDA team members planted light sensors a kilometer
below the surface of the ice sheet, shielding the sensors from
stray particles. But that is not all they use the ice for; it is their
detector as well.

Other neutrino detectors need giant cylinders or spheres
of water, or vast chunks of metal—they are the mass that the
neutrinos interact with, and since neutrinos are so insensitive
to mass, a detector has to have an enormous amount of it to be
able to detect any neutrinos. Luckily for the AMANDA project, there's one thing that they don't have a shortage of: ice.
Tons and tons of Antarctic ice. All this ice acts as an enormous detector. When a weakly interacting particle penetrates
deep into the ice, there is a chance it will interact with the ice
itself, setting off a flash of light that is, in turn, picked up by
the buried sensors. In 2000 the AMANDA team upgraded its
equipment, and members are now gathering data every
Antarctic summer. They hope to catch WIMPs while they
search for neutrino sources in deep space.

The National Science Foundation plans yet another experiment, known as IceCube, that will span a cubic kilometer
of Antarctic ice. It will be like having a detector that weighs a
billion tons, about the same as ten thousand aircraft carriers.

It's a WIMP trap that boggles the imagination. It is hard to believe that exotic dark matter will be able to escape detection for much longer.

Earthbound experiments are closing in on exotic dark matter, and even in deep space, far from any source of light or object that might reveal its presence, dark matter cannot hide. Scientists are now glimpsing the primordial "filaments" that collapsed into galaxies and galaxy clusters. They are also seeing hints of the ubiquitous hydrogen "fog" that made the universe a very dark place for the first hundred million years of its existence.

Unlike a WIMP, which feels only the effect of the weak force, an atom of hydrogen is affected by the electromagnetic force and its carrier, the photon. A hydrogen atom with an electron, for example, cannot resist photons of certain energies. The hydrogen slurps them up faster than a six-year-old wolfs down an ice-cream sundae. Just as six-year-olds prefer certain flavors of ice cream, hydrogen—like other atoms—absorbs only certain colors of light from the rainbow of the spectrum. If you shine a beam of white light through a cloud of hydrogen gas and then split the light with a spectrum, you will see that the rainbow is pockmarked by dark lines at the frequencies where hydrogen absorbs light. Because of this absorption, light from a distant object, such as a quasar, that passes through a cloud of neutral hydrogen and helium will be scarred with lots of lines. Making things more complicated, those lines can be shifted by the Doppler effect, so if the cloud is moving (which it almost certainly is), those lines shift toward the red or the blue part of the spectrum depending on whether it is moving away or toward Earth. (This is on top of the redshift due to the Hubble expansion.) Worse still, any light that is coming from a very distant source is almost certainly going to pass through oodles of these clouds, each of which is moving in a different direction and at a different speed. This means that each cloud will stamp its unique set of dark lines upon the light coming from a distant quasar. Astronomers have long known that light from quasars is riddled with a veritable forest of black lines in the spectrum: it is known as the Lyman-alpha forest, after the most important absorption line.

The Lyman-alpha forest was discovered in the early 1970s, and since the mid-1990s scientists have been successfully reconstructing the shape of the intergalactic gas pockets by analyzing the pattern of lines. All the data seem to support the cosmologists' picture of the early universe: clumps of matter in the early universe coalesced into filaments. These filaments formed galaxies and galaxy clusters, whereas the rest of the universe, the space between the filaments, is vast, empty wasteland. Thanks to the Lyman-alpha and other spectral lines, astronomers are beginning to spot these filaments directly. Just as hydrogen gas absorbs light, under certain conditions it will emit the same wavelengths of light. In 2001, astronomers at the European Southern Observatory claimed that faint Lyman-alpha emissions gave them a direct view of one of these filaments. They are spotting the faint signal from a cloud of gas deep in the middle of darkest intergalactic space. Even the darkest objects can now be seen.

What's more, scientists are beginning to get their first view of the darkest ages of the universe. During recombination, 400,000 years after the big bang, electrons settled down and bound themselves to hydrogen and helium nuclei. As the first stars ignited, sputtering into life, much of their light was absorbed by the neutral hydrogen. But once enough stars, galaxies, and quasars had formed, the combined light from all those objects stripped the electrons from the atoms and the fog of hydrogen could no longer absorb light. It was the dawn after the cosmic dark age: the era of reionization.

In mid-2001 scientists from the Sloan Digital Sky Survey claimed that they had seen the first hints of the hydrogen fog remnants at the end of the dark age. Light that left a quasar only 900 million years after the big bang showed a dark swath in its spectrum that signified absorption by a hydrogen cloud. It is a relic from the time when reionization was nearly complete. As scientists find ever more distant quasars, they will be able to push back into the depth of the dark age, spotting dark matter in a dark universe.

With gravitational lenses, x-ray telescopes, WIMP traps, Lyman-alpha observations, and the signature of the cosmic dark age, astronomers are finding all the hiding places of dark matter, both baryonic and exotic. If just a few of these techniques pay off, as they are beginning to do, astronomers will pin down where dark matter is sequestered; they will understand the secrets of Ω_m. They will have conquered the invisible.

Yet mass is merely half of the mystery behind Ω, because Ω has two components: mass, represented by Ω_m, and energy, represented by Ω_Λ. Cosmological observations indicate that Ω_m represents 35 percent of the stuff in the universe; 5 percent is baryonic, and the remaining 30 percent, exotic. When scientists understand Ω_m, they will have figured out precisely what that 35 percent is made of, where it resides, and how it behaves.

That would leave only Ω_Λ unaccounted for; Λ is shorthand for the cosmological constant, the bizarre repulsive force that is driving the universe apart, and Ω_Λ is its contribution to the "stuff" in the universe. The nature of the cosmological constant is the newest cosmological quandary; it is a puzzle that was posed at the beginning of the third cosmological revolution when the supernova data showed that the universe's expansion, instead of slowing down, is speeding up. It is the wilderness where scientists are most lost, but here too there is hope. Oddly enough, the answer seems to be hidden in the vacuum of space.

Chapter 12
The Deepest Mystery in Physics

[Λ, THE VACUUM, AND INFLATION]

I believe that the vacuum, being the state in which all possible physical phenomena are present, in a virtual way, but still present, will win the record for the highest complexity.

— CARLO RUBBIA

The vacuum is the most complex substance in the universe. Within it are all particles and all forces, even those unknown to science. Physicists now believe that the vacuum—the emptiness in deep space, or even in a vacuum chamber—holds the secret to the newest question in cosmology: what is this mysterious Λ, this antigravity force that flattens out the universe and pushes galaxies apart? A mere decade ago, Λ was a mathematical absurdity. Today, it is a very real force that gnaws at the heart of cosmology.

The very ridiculousness of Λ makes it very difficult to understand; within the space of a few years, this dark energy has rewritten cosmologists' understanding of the universe. But once scientists figure out what Λ really is, they will have unraveled the deepest mystery in physics today. Not only will they understand the dark energy, they will understand the physics that drove the big bang itself. They will be able to see beyond even the era of quark-gluon plasma, to a time a billionth of a trillionth of a trillionth of a second after the big bang, to a time when the quantum vacuum held the fate of the universe in its grasp.

It seems like a contradiction to say that the vacuum is the most complex phenomenon in the universe. The very definition of the vacuum is the absence of everything, a space filled with nothing at all. In the 1930s, though, quantum physicists discovered, much to their surprise, that the vacuum isn't ever truly empty. It is seething with activity, filled to the brim with particles and energy.

This silly-sounding notion comes from one of the fundamental ideas in quantum mechanics: the Heisenberg uncertainty principle. In the mid-1920s, German physicist Werner Heisenberg was figuring out the mathematics that govern the subatomic world. Those mathematical rules, the newly minted laws of quantum mechanics, had a shocking consequence that even Heisenberg did not expect: there are some things that scientists just cannot know, no matter how hard they try. More precisely, there is a fundamental relationship between certain attributes of a particle, like its position in space and its momentum (how much "push" the particle has). The more you know about a particle's momentum, the less you know about its position; conversely, the more you know about a particle's position, the less you know about its momentum. As you gain knowledge about one of these properties, you lose knowledge about the other.

In a sense, these connected properties are like a lump of Play-Doh; if you try to flatten the lump in one direction, the rest of the blob squirts out in the other direction. When you reduce the uncertainty about one of a particle's properties, the uncertainty in the other, connected property must get larger to compensate. This is the essence of the Heisenberg uncertainty principle.

Most people describe the Heisenberg uncertainty principle as a phenomenon of measurement. If you bounce a photon off an electron to mark its position, you give it an unknown kick of momentum, messing up your knowledge of the particle's current momentum. While this is true, the uncertainty principle is more profound than that; it works even when nobody's measuring anything. It is a fundamental law about the way the universe works, whether somebody is observing a subatomic particle or not. Nature herself must obey the dictates of Heisenberg's principle.

Two sets of Heisenberg-linked properties are particularly important to understanding the quantum vacuum. Momentum and position have already been mentioned; if you know a lot about how much "push" a particle has, then you know very little about where it is, and vice versa. Energy and time are also linked in this way: the more you know about how much energy a particle has, the less you know about *when* it has that energy. These uncertainty relations have a very strange consequence. They fill the vacuum with infinite numbers of evanescent particles that blink in and out of existence.

Imagine a tiny box in a vacuum chamber. Everything inside that box has a well-known position; after all, it is in a small box. However, that certainty about position means that the contents of the box cannot have a well-known momentum. If the box were truly empty, the stuff inside would have zero momentum; after all, the absence of particles can't push very hard. And if the stuff inside the box had zero momentum, you would know *precisely* how much momentum the stuff inside the box has—a violation of the Heisenberg uncertainty principle because you already have lots of information about its position. Thus, the principle states that you cannot know what's inside the box! This is naturally explained by the energy-time uncertainty; on very small time scales, particles are constantly blinking in and out of existence, so at any given moment, you don't know how much energy is inside the box. It's a startling concept—that the subatomic world is constantly seething with particles that appear out of nowhere and promptly disappear again. But it is true; the Heisenberg uncertainty principle forces nature to create and destroy these particles constantly, at all points in space, even in the deepest vacuum. The smaller the box, the worse the problem: the less you must know about the momentum inside the box, even though it is filled with the vacuum.

Quantum physicists are forced to conclude that the vacuum isn't truly empty. It is seething with particles and energy. On a smaller and smaller scale, the particles that appear and disappear will have more and more momentum; they will have more energy (and be more massive) and live for a shorter time, thanks to the energy-time relationship. So, on relatively large scales, lightweight particles like electrons and antielectrons are constantly popping in and out of existence, but on smaller and smaller scales, heavier particles like muons and taus (and undiscovered, massive particles, like WIMPs and other sparticles) become more and more important.[1]

This is not just a physicist's fantasy. Scientists have watched these evanescent particles push metal plates around. In 1996, physicist Steven Lamoreaux, then at the Los Alamos National Laboratory, measured that push, a phenomenon known as the Casimir effect, named after the Dutch physicist who predicted it. The force involved in the Lamoreaux experiment was very tiny—if you chopped up an ant into about thirty

---

1. This principle is part of the reason that more massive particles tend to be less stable. The relationship between energy and time means that higher-energy phenomena (like more massive particles) tend to appear on shorter time scales. This is also why the searches for the new physics that take place on tinier and tinier scales require more-energetic particle accelerators. The more energy you put into a collision, the greater the uncertainty in energy and momentum there can be (if you throw a ball at 100 mph, you have a greater possible range of speeds than if you can only throw it at 50 mph), so you can see objects on a smaller and smaller scale. If you make that scale small enough, you might see supersymmetric particles and other undiscovered phenomena.

thousand pieces, it was roughly equivalent to the weight of one slice — but the force of the evanescent particles was clearly there. Since then, several other teams of physicists have measured that push with different pieces of equipment. The vacuum really is filled with particles and with energy.

This *zero-point energy* of the seething quantum vacuum also affects the interactions between particles. When physicists calculate the properties of any interaction, they must correct their estimates to account for the influence of the innumerable particles that constantly pop in and out of the vacuum. Even though the number of possible interactions is technically infinite, as is the energy in the vacuum, scientists can safely ignore all but the largest without messing up their calculations too much.[2] However, if any particles are missing from those calculations, such as an undiscovered supersymmetric particle, then there will be a discrepancy. A very precise experiment might yield results slightly different from what is calculated by theoreticians. This is why scientists get excited when a credible, carefully performed experiment — such as the CERN discovery of too many tau particles or the Brookhaven measurement of the magnetic moment of the muon — disagrees with theoreticians' predictions. It might be the signature of a particle that has not been properly accounted for, a particle unknown to science. Such a particle would be floating about in the vacuum, because everything is in the vacuum.

Empty space is an incredibly complex substance, and scientists are just beginning to understand its properties. The need is more pressing than ever before, because they think

2. The method for accounting for the infinities is called renormalization, and renormalization in quantum electrodynamics was part of the work that won Richard Feynman, Julian Schwinger, and Sin-Itiro Tomonaga the 1965 Nobel Prize. A decade later, physicists Gerardus 't Hooft and Martinus Veltman figured out how to renormalize electroweak forces, but their technique only worked if theorists added a then-undiscovered top quark, so physicists started looking for it and found it in 1995. The discovery of the top quark was a major victory for the 't Hooft-Veltman theory, and netted the discoverers the 1999 Nobel Prize.

that the energy of the vacuum, the zero-point energy that is everywhere in the universe, is forcing the universe apart.

The discovery of this strange antigravity force has brought cosmology back to the confusion of the 1920s, when Einstein grappled with the idea that the universe was unstable. To deal with this calamity, Einstein altered his equations by putting in a fudge factor, Λ, that counteracted the effects of gravity. Just as the outward pressure of fusion energy counteracts the sun's gravitational collapse, the outward "pressure" of Λ counteracted the attractive forces between galaxies and galaxy clusters, keeping the universe in a stable equilibrium. There was no experimental support for Λ, no physical reason to believe that some sort of antigravity force existed, so when Hubble discovered the expansion of the universe, Einstein hastily withdrew the idea and later dubbed it the greatest blunder of his career. For seven decades, Λ was left on the heap of discarded and discredited ideas. It simply didn't fit into the way cosmologists thought the universe was put together.

That changed suddenly and dramatically in 1998 when the supernova hunters discovered that the universe's expansion was accelerating rather than decelerating. Even though nobody had suspected that an antigravity force was at work counteracting the gravitational pull of galaxy clusters, there had to be one, just as surely as a baseball speeding at ever increasing speeds into the air must be propelled by some unseen force that opposes the pull of gravity. All the measurements pouring in—of cosmic background radiation, galaxy distribution, supernovae, big bang nucleosynthesis—supported the bizarre conclusion that a mysterious antigravity force, dark energy, must make up 65 percent of the "stuff" out there; Λ was back with a vengeance. But what could this dark energy be? The quantum vacuum might hold the answer.

Physicists are in the very early phases of investigating the properties of dark energy. They don't even know its basic properties. For instance, scientists don't know whether it has

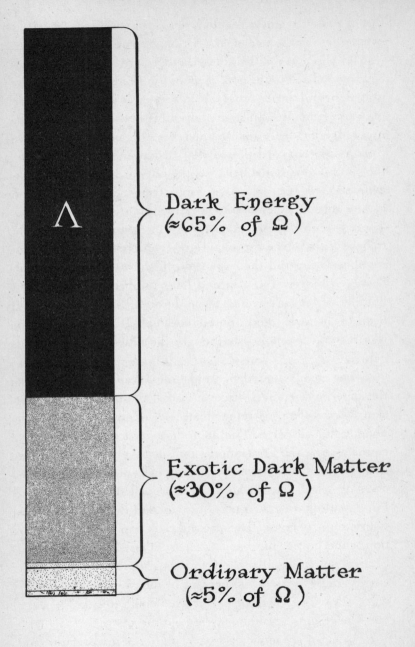

Λ

Dark Energy
(≈65% of Ω)

Exotic Dark Matter
(≈30% of Ω)

Ordinary Matter
(≈5% of Ω)

remained constant throughout the life of the universe—that is, whether the cosmological constant is truly constant—or whether it changes strength over time. (One popular model looks like a time-varying version of the cosmological constant, but it requires some sort of new particle or field in the universe called *quintessence*, named for the fifth ancient element of the universe after earth, air, fire, and water.) At this point, scientists simply don't know whether the time-varying model or constant model of dark energy is correct, and they don't know what the source of the cosmological constant or quintessence is. Nevertheless, a leading contender for the source of dark energy is the energy latent in the vacuum.[3] If the particles that wink in and out of existence exert a pressure that can push around metal plates, it seems reasonable to think that they are pushing apart galaxy clusters too. Physicists don't have a detailed mechanism for how this works yet; worse, the current standard-model calculations imply that the energy in the vacuum is so great that galaxies should be thrown apart at tremendous speeds, much greater than what astronomers observe. There is *much* too much energy in the vacuum to explain Λ, at least according to the standard model.[4] Nonetheless, there is hope that when physicists refine the standard model, they will make better calculations of the processes that go on in the vacuum and explain the difference.

For years, physicists have been trying to extend the standard model in an attempt to unify the strong, electroweak, and gravitational forces. The most promising theory is known as M-theory.[5] M-theory is a supersymmetric theory; that is, if

3. The other contender is quintessence's mystery particle that exerts a repulsive force, and so quintessence's dark energy is not the vacuum energy. However, the quintessence models seem to have as many troubles, if not more, as those based on the cosmological constant, which use the vacuum energy as the source of dark energy.
4. How much too much? Up to 120 orders of magnitude. This is a mindboggling number. For instance, the difference between the mass of a single atom and the mass of all the atoms in the universe does not approach 120 orders of magnitude.
5. M-theory is the extension of the better-known superstring theories and brings them all under the same mathematical banner. It will be discussed in chapter 14.

M-theory is right, then one version of supersymmetry must be as well. If it is correct, M-theory would explain the forces of nature at incredibly high energies, in incredibly small spaces, and over very short timescales. In other words, M-theory might explain the processes that give the vacuum its energy. But those processes are not the only things that are hot, tiny, and short-lived. So is the big bang. Solving the mystery of the vacuum may give scientists a direct view of the physics of the big bang and the period almost immediately following it: inflation.

The concept of inflation is an important part of modern big bang theory. Physicists think that, before the formation of hydrogen and helium, before protons and neutrons condensed out of the quark-gluon plasma, the universe, in a moment, expanded at an enormous speed. This period of inflation is very troublesome—it requires a new sort of physics to explain it—but it solves two of the nagging problems that physicists had about the nature of the universe: the horizon problem and the flatness problem.

We have already encountered the effects of the horizon problem in chapter 5, with respect to the cosmic background radiation. The problem stems from the fact that only about 400,000 years passed between the big bang and the era of recombination. Since information travels at the speed of light, any given atom of hydrogen could only feel the influence, the gravitational pull or the radiation, of another atom within 400,000 light-years. That is, when the era of recombination ends, an atom cannot be "causally connected" to another atom that is more than 400,000 light-years away. Regions about 400,000 light-years across are the biggest causally connected regions in the early universe, so they are the largest regions that collapse under their own gravity. This phenomenon leads to a maximum size of hot spots in the cosmic background radiation. But we also know that the cosmic background radiation is about 2.7 degrees in all regions of the sky, give or take a few millionths of a degree. How can there be such a remarkable similarity among all the regions of the sky if they are not causally connected with one another?

Imagine that every time Captain Kirk visited an undiscovered alien civilization on the *Enterprise*, every alien creature who was beamed up was always wearing a green turtleneck with red trousers. No matter where in the universe the *Enterprise* goes, aliens always wear green turtlenecks with red trousers, even though those civilizations have been living in isolation, with no communication with their neighbors. It couldn't be a coincidence, so there has to be some underlying cause, something that impelled the alien civilizations to wind up wearing green turtlenecks. Maybe they all came from the same home planet millions of years before, but it is almost inconceivable that they all developed the same dress code totally independently. This scenario is analogous to the horizon problem. How could regions of the sky that are causally disconnected develop in the same way? How can they have the same temperatures, pressures, and densities, and generally look alike, even though they shouldn't have been able to influence one another? The incredible uniformity of the heavens seemed too implausible to believe, and scientists did not have an explanation.

We have also encountered the other problem with big bang theory, the flatness problem, albeit indirectly. Scientists had long suspected that the universe was fairly close to flat. They didn't know quite how close, but they knew it was reasonably flat, since they didn't see any obvious distortions of spacetime due to curvature. However, it is extremely unlikely that the universe should have such a near-flat geometry. Indeed, now that we know for sure that the universe is almost perfectly flat, it is vanishingly unlikely. If you were to choose a "random" closed universe, it would tend to explode, collapse, and die in less than a trillionth of a trillionth of a second. A "random" open universe would have very little matter; everything would fly apart very quickly, and the universe would be quite saddle-shaped rather than almost flat. A nearly flat universe was a seemingly improbable event, something like a monkey sitting down in front of a typewriter and banging out *Ulysses* on his first try.

The flatness and horizon problems gave cosmologists heartburn. Though the big bang theory explained much about the nature of the universe and predicted the existence of the cosmic background radiation, it did nothing to explain these cosmic coincidences.

In 1980 physicist Alan Guth of Stanford University proposed a way to solve both problems with one theory: inflation. Inflationary theory states that for a very short time, the universe underwent a superrapid expansion. The size of the universe doubled and doubled and doubled again; the rate of the inflation was so great that the fabric of spacetime was blown apart at a rate faster than the speed of light.[6] It would blow up from a fraction of the size of a neutron to the size of the visible universe in a tiny, tiny amount of time. But this rapid inflation was unstable, and it shut down after about 10^{-32} seconds. This dramatic expansion, as short-lived as it was, had two important consequences.

First, it solved the horizon problem. Right after the big bang happened, everything in the supertiny universe was able to affect everything else; all the energy in the universe could spread out pretty much uniformly. Everything in the cosmos was causally connected in the first few moments of creation. As soon as inflation kicked in, though, spacetime expanded so rapidly that regions were blown away from one another, effectively at speeds faster than the speed of light. When inflation stopped, these regions were so far apart that it looked like they could not have had any causal contact with one another. Yet, in the preinflationary period, they *were* in causal contact, so it is not a coincidence that the regions look

6. This does not violate Einstein's speed-of-light limit to information transfer. Einstein's laws restrict the speeds of things that travel *along* the fabric of spacetime; the inflation of the universe says that the fabric itself, which isn't subject to Einstein's speed limit, expanded faster than the speed of light.

similar. However, they are not too similar. Thanks to the Heisenberg uncertainty principle, there had to be tiny quantum fluctuations that disturbed each region; when inflation stopped, each causally disconnected region, altered slightly by those fluctuations, evolved independently. After the end of inflation, those differences in the distribution of mass and energy led to the hot and cold spots in the cosmic background radiation. In fact, the inflation model predicts that those fluctuations would have a property known as scale invariance,[7] precisely what other theories about the cosmic background radiation predicted, and what recent observations of the microwave background and of galaxy-cluster distributions are showing.

Second, inflation solves the flatness problem. No matter what shape the universe started out as, inflation blew it flat, just as a wrinkled balloon becomes smooth as you inflate it, and as it gets bigger and bigger, the surface looks flatter and flatter. Such a rapid and dramatic inflation would make the universe almost perfectly flat, give or take a tiny bit of curvature, all but undetectable to astronomers. Instead of relying upon a cosmic coincidence to have a flat universe, Guth's inflation theory gave a reason why the universe had to be flat. But what could drive this dramatic inflation, and how could it suddenly shut off? Here too the answer seems to lie with the vacuum.

---

7. When scientists analyze fluctuations, they look at a *spectrum*. Just as you can figure out the components of a beam of light—what colors it contains—by passing it through a prism, you can figure out the components of fluctuations or noise—what size the fluctuations tend to be—by a mathematical trick. The result is something that reveals how frequent the fluctuations of a certain size are; for example, for every 100 fluctuations that are 1 meter "wide," you might get 10 fluctuations that are 10 meters across, and 1 fluctuation that is 100 meters across. This particular distribution, in which the size is inversely proportional to how common a fluctuation is, is called a scale-invariant spectrum, because regardless of what measurement of length you use—meters, feet, or furlongs—the shape of the curve that describes the fluctuations will look precisely the same. In the 1970s, Edward Harrison, Yacov Zel'dovich, P. J. E. Peebles, and J. T. Yu showed that the cosmic background radiation should have a scale-invariant spectrum, also known as a Harrison-Zel'dovich spectrum. Alan Guth's inflationary scenario led automatically to a scale-invariant spectrum for mass fluctuations, adding support to inflationary theory.

Thanks to the zero-point energy, the particles constantly winking in and out of existence, the vacuum has a "pressure" of sorts. If the zero-point energy was greater in the early universe's vacuum than it is in today's vacuum, the pressure of the vacuum would have been immensely greater then than it is today. This enormously enhanced energy of the vacuum would try to expand in all directions, inflating the fabric of spacetime with tremendous speed and power, smoothing out the lumpiness of the universe as it goes. However, the higher energy state of the early, primordial vacuum, which scientists term the *false* vacuum, makes it unstable. It can only last for a short time. In less than a millionth of a millionth of a millionth of a millionth of a second, the false vacuum would collapse. It would condense like steam into water, reverting into the "true," modern-day vacuum that has less zero-point energy and a much smaller pressure. The supervacuum would become the vacuum that we see today.

The transition from the false vacuum to the true vacuum would be violent and sudden, releasing an enormous amount of energy. Once the transition occurs at any given point in space, it would set up an enormous spherical shock wave traveling at the speed of light in all directions; inside the giant bubble that forms as this shock wave expands, the false vacuum condenses into the true vacuum. Our observable universe, with its true vacuum, presumably resides within one of those bubbles (or several that expanded into one another, linking themselves together.) There might be other bubble universes out there, invisible to our instruments, separated from our own bubble universe by a wall of false vacuum.[8]

---

8. It may be that our true vacuum isn't really the lowest energy state; there may yet be a "truer" vacuum that contains less zero-point energy than our own. In 1983, two scientists published a paper in *Nature* that explored what would happen if physicists accidentally triggered a second round of condensation from our vacuum to a still lower energy state. To make a long story short, our universe is destroyed. This was a basis for the protests at Brookhaven National Laboratory mentioned in chapter 8; footnote 7 there explains the universe's continued existence.

The concept of inflation is mindboggling, but the mathematics of it works out and it solves the horizon and the flatness problems. Better yet, the force that drives inflation might be related to the recently discovered dark energy that suffuses the universe. Perhaps the force that gently pushes the galaxies away from one another was stronger in the past. The very weirdness of Λ makes inflation look reasonable. So does the cosmic background radiation.

Since 1980, scientists had proposed a number of alternatives to inflation. The most promising of them relied upon *topological defects*, irregularities in spacetime itself caused by exotica such as cosmic strings or magnetic monopoles. Though the details of each type of defect were different, the overall effect of these topological defects was pretty much the same, explaining how the early universe developed its present-day structure. However, there was one substantial difference. Topological defects would produce a cosmic microwave background spectrum different from that produced by inflation.

According to the theory of inflation, all the pockets of matter and energy were inflated rapidly, and the quantum fluctuations of the vacuum—tiny ripples in the distribution of that energy—expanded with spacetime to form large fluctuations. As soon as inflation shut off, those fluctuations began contracting all at the same time. This was the start of the acoustic oscillations in the primordial universe. Because all these regions began contracting at the same time, the one-degree hot spots all reached their peak temperature simultaneously; they were locked in phase, even though they could not communicate with one another. This is what gives the spectrum of the acoustic oscillations a hump-and-dip appearance, with numerous narrow peaks and valleys. With topological defects, on the other hand, all the fluctuations do not march in step, so some of the one-degree features would be reaching their peak temperatures as others reached their lowest ones. This would translate to a large, smeared-out peak in the spectrum, rather than lots of little narrow ones. When Boomerang returned with its first results, it was the end of the topological models. "With topological defects, you only predict one peak, but a very broad one. This peak is way too narrow," said the University of Pennsylvania's Max Tegmark. "This really means that most of the rivals to the standard theory just died."

Even though it was a snapshot of a time 400,000 years after inflation ended, the cosmic background radiation gave scientists a view of the very earliest moments after the big bang. For a year, after the death of the topological defects alternative, it looked as though inflation was the only game in town. Then, in 2001, cosmologist Paul Steinhardt of Princeton University and his colleagues came up with a new scenario, one that was as powerful as inflation, explaining the horizon and the flatness problem, and yet had very different underpinnings. Instead of a big bang, Steinhardt argued, the universe was born in a big splat.

At first glance, the new model, based on the physics of M-theory, seems surreal. It takes place in eleven dimensions, six of which are rolled up and can safely be ignored. In that effectively five-dimensional space float two perfectly flat four-dimensional membranes, like sheets drying on parallel clotheslines. One of the sheets is our universe; the other is a "hidden" parallel universe. In the latest version of the theory, our unseen companion slowly floats toward our universe. As it moves, it flattens out—although quantum fluctuations wrinkle its surface somewhat—and gently accelerates toward our membrane. The floater speeds up and splats into our universe, whereupon some of the energy of the collision becomes the energy and matter that make up our cosmos. Because both membranes are roughly flat, our postcollision universe remains flat as well. "Flat plus flat equals flat," says Steinhardt.

Because the membrane floats so slowly, it has a chance to equilibrate, giving it more or less the same properties over its entire surface, although the quantum fluctuations cause some irregularities. That explains why our universe looks roughly (but not exactly) alike in every direction. The slow motion of the membrane solves the horizon problem. But whereas inflation solves the horizon and flatness problems by a quick, violent process, Steinhardt points out that "this model works in the opposite sense: slowly, but over a long period of time." Another attractive feature is that it gets rid of the puzzling singularity at the beginning of the universe: instead of a pointlike big bang, the universe is formed in a platelike splash. To later observers, the big bang and big splat would be almost indistinguishable, since the big splat looks like a big bang universe right after inflation shut down; everything from 10^{-32} seconds onward would be almost the same—the formation of protons and neutrons, the creation of hydrogen and helium, and the recombination that released the cosmic background radiation.

Though this idea is very new and has not been fully digested by the scientific community, it excites scientists because it links the beginning of the universe to the increasingly compelling ideas behind M-theory. "It's the first really intriguing connection between M-theory and cosmology," says Princeton physicist David Spergel. "This is sort of an ur–big bang." The big splat theory proves that the basic ideas of M-theory can lead to a consistent picture of the universe that explains the cosmos about as well as inflation does. If the model is right, however, it could have some very nasty consequences. The model's official name, the *ekpyrotic* scenario, Steinhardt explains, comes from the Stoic term for a universe periodically consumed in fire. That is appropriate because at any moment the unseen membrane could be floating toward us, toward another collision. Indeed, Steinhardt says, we might have already seen the signs of impending doom. "Maybe the acceleration of the expansion of the universe is a precursor of such a collision," he says. "It is not a pleasant thought."

At the moment, the inflationary and ekpyrotic scenarios are the two models for the beginning of the universe; whichever is correct will take our understanding to the first tiny, tiny fraction of a second after the big bang, and maybe even earlier. But to tell which model is correct means that scientists must probe the very earliest moments of the universe; they have to see beyond the walls of fire that surround us and figure out precisely what was happening in the first few moments after the big bang. It is a daunting task, but the instruments that can do it have already been built. Even now, these brand-new laboratories are picking up the signals from the beginning of the universe. They are looking for wrinkles in time.

Chapter 13
Wrinkles in Spacetime

[GRAVITATIONAL WAVES AND

THE EARLY UNIVERSE]

Every wave is wealth to Daedalus,

Wealth to the cunning artist who can work

This matchless strength.

— RALPH WALDO EMERSON, "SEASHORE"

Gravitational waves are a way of looking beyond the walls of fire, the cosmic background radiation that surrounds us in every direction. We have already seen how the cosmic background radiation is the mainstay of modern cosmology — it confirmed the big bang theory, revealed the shape of the cosmos, helped determine the amount of dark matter and dark energy that populates the universe, supports the theory of inflation (and its near equivalent, the ekpyrotic universe), and saps the strength of alternative theories that need exotica like topological defects. But at the same time, the cosmic background radiation is also an impediment, a barrier that prevents astronomers from seeing the very earliest ages of creation.

The seething ball of plasma that permeated the cosmos emitted radiation when it finally cooled enough, and the fiery walls of ancient plasma surrounding us in every direction became transparent. Since that plasma was opaque to light, none of the photons from the era of inflation or from the big bang survived to the present day. They were all absorbed by the plasma, their information scattered and dissipated in countless directions. Astronomers see none of the light created during the first 400,000 years after the big bang. The walls of plasma are the limit of their vision. "This is probably the farthest light that can be observed," says Phil Mauskopf, an astrophysicist at the University of Wales in Cardiff.

To see beyond the plasma walls, scientists are trying to find another signal from the early universe, a signal that would not be destroyed by the opaque plasma. They are looking for gravitational waves, ripples in space and time that rattle through the universe at the speed of light. Physicists know they exist and have seen their effects. They are looking for the waves' signatures in the cosmic background radiation, and any day now, they will measure one as it squashes the Earth, distorting the very fabric of spacetime. The first sign of gravitational waves came from Little Green Men.

In 1967, Jocelyn Bell, a graduate student at Cambridge University, found an object that blinked on and off in the sky, emitting incredibly regular pulses of radio waves like some sort of cosmic beacon. At first, the mysterious object was jokingly dubbed an LGM, after the Little Green Men who seemed to be trying to send us a message. But as other astronomers all over the world turned their radio telescopes to the sky, they spotted several similar objects and ruled out an artificial origin. Bell hadn't discovered little green men; she had seen the first *pulsar.* (In 1974, her adviser, Anthony Hewish, was awarded the Nobel Prize for the discovery.)

A pulsar is a neutron star, the burned-out husk of a medium-mass star. As it spins, it emits a powerful beam of radiation from its magnetic poles. That beam sweeps across the sky in a cone. Anyone illuminated by that beam would see the star blink on and off, just as anyone in the spinning beam of a lighthouse sees the lighthouse flashing. A pulsar is a stellar lighthouse in deep space.

Barely a month before Hewish won the Nobel Prize, Joseph Taylor, then an astronomer at the University of Massachusetts, Amherst, and his graduate student Russell Hulse discovered a new type of pulsar. Its bursts seemed to be less regular, speeding up and slowing down, rather than ticking away with an unchanging tempo like a metronome. Hulse and Taylor realized that they had discovered a binary pulsar, a pulsar orbiting an unseen companion. As the pulsar sweeps out its orbit in space, it zooms toward and away from the Earth, making the pulses seem to speed up and slow down, even though the star itself spins with clocklike precision. The pulsar, ticking away with a steady beat as it orbited its companion, let Taylor and Hulse test one of Einstein's predictions—gravitational waves—for the first time.

Einstein's rubber-sheet picture of spacetime explained the source of gravitational attraction as a warp, a dimple, in the fabric of space and time. As we have seen, this idea was phenomenally successful; it predicted gravitational lensing, which Eddington spotted in 1919, as well as the irregular orbit of Mercury, the explanation of which had eluded Newtonian scientists for three centuries. But even as Eddington and Einstein confirmed these effects of general relativity, the theory predicted several other phenomena that were untestable at the time. Gravitational radiation was one of them.[1]

Gravitational waves are a direct consequence of the rubber-sheet model of general relativity. Just as a massive object

1. Another one, the Lens-Thirring effect, also has to do with the effect of moving bodies on the fabric of spacetime. Einstein's theory predicts that a spinning massive object, like the Earth, "drags" the fabric along with it, twisting it the way a fitful sleeper twists the sheets around himself. This frame-dragging effect has only recently been detected around black holes and neutron stars. A very expensive satellite, Gravity Probe B, will look for the effect as it orbits the Earth.

sitting on the fabric of spacetime creates a dimple, moving objects, under certain conditions, create wrinkles in the fabric. Those wrinkles, tiny distortions in space and time, zoom away at the speed of light. These waves also carry energy; anything emitting a gravitational wave will lose a tiny bit of its speed.

The 1974 binary pulsar gave scientists a way to test this prediction for the first time. According to general relativity, the pulsar and its unseen companion must emit gravitational waves as they dance around each other. Those gravitational waves carry away some of the stars' energy, slowing them down and causing them to fall inward. Their orbits get shorter and shorter as the two stars get closer and closer together. In 1978, Taylor and his colleagues showed that this was precisely what the binary pulsar system was doing. Every year, the pair's orbit was seventy-five milliseconds shorter than it was the previous year. This is a tiny amount, but thanks to the orbiting metronome of the pulsar, it was not such a difficult measurement to make. This was the first evidence for gravitational waves. In 1993, Taylor (*and* his former student, Russell Hulse) won the Nobel Prize for the discovery.[2]

The theory of relativity states that lots of objects and events must radiate energy in the form of gravitational waves. Massive stars orbiting each other, black holes swallowing a star-size lump of matter—all these create gravitational ripples. So did inflation. As the fabric of the universe expanded with a sudden burst of energy, some of the energy wrinkled that fabric, causing gravitational waves. Unlike photons from that era, the gravitational waves created during inflation and its aftermath would not be scattered or absorbed by the ubiquitous plasma. If scientists had an instrument that could measure the passing of gravitational waves, they might be able to pick up signals that came directly from the birth of the universe. Unfortunately, they didn't have any instrument sensi-

2. Joseph Taylor invited Jocelyn Bell (now Jocelyn Bell Burnell) to accompany him to the Nobel ceremony.

tive enough to detect the minute distortions of space and time caused by gravitational waves. Until now.

In October, 2000, an enormous L-shaped laser facility in the state of Washington began its search for gravitational waves. Shortly thereafter, a nearly identical laboratory opened up in the Louisiana bayou. Together, the two gigantic instruments make up LIGO, the Laser Interferometer Gravitational-Wave Observatory, a nearly $400 million project to spot gravitational waves directly for the first time.

The equations of general relativity describe what a ripple in spacetime looks like. It travels at the speed of light, and it has a peculiar shape; as it stretches spacetime in one direction, it squashes it in the other. If you took two yardsticks and held them at right angles as a gravitational wave gurgled through, you would see one yardstick contract and the other expand. If you constantly compared the sizes of the two sticks, in theory you could spot a gravitational wave when one stick suddenly became larger than the other. However, this effect is incredibly subtle. Even if you have mile-long yardsticks, the difference in length due to a passing gravitational wave would be less than a tiny fraction of the size of a proton. Measuring that minuscule effect is a daunting engineering challenge. For decades, scientists didn't know how to design yardsticks and measuring devices precise enough to spot it.

The two facilities of LIGO are the most sophisticated and expensive yardsticks in the world. Both LIGO facilities work on the same principle, exploiting the wavelike properties of light to make a pair of superaccurate measuring devices. Each facility is an enormous interferometer.[3]

Interferometry relies upon light's wavelike nature. Under certain conditions, you can think of light as a series of ripples, something like ocean waves. These ripples have crests and

3. We have already encountered the idea of interferometry in describing the DASI and CBI microwave telescopes; they used the wavelike properties of light to give them a powerful, steerable antenna, just as modern missile cruisers do. The underlying principle here is the same, though the end result is somewhat different.

troughs, just like water waves; indeed, scientists talk about the wavelength of light—the distance between successive crests—when they want to describe light's color. (The longer the wavelength, the redder and less energetic it is.)

Imagine a wave of light rippling through space. All of a sudden, it is divided in two by a beam splitter. (A lightly silvered mirror that lets half the light through will do the trick.) The two beams travel independently for a while, and then they get recombined at a target some distance away. The two beams start off marching in lockstep—a crest in one moves parallel to a crest in the other. In scientific jargon, they are "in phase." If the two beams make a journey of the same length, the crests that leave the beam splitter at the same time will also reach the target at the same time. Crests will meet crests; the two will reinforce each other. Two beams, in phase, become a single bright beam when they are combined. But if the paths are different lengths, the story gets more complicated. If path A is a hair longer than path B, then it will take a fraction of a second longer for a crest moving along path A to reach the target than one moving along path B. When a crest from beam A reaches the target, the crest of beam B has already traveled on ahead. The two beams are now out of phase. In the special case where path A is exactly half a wavelength longer than path B, you can see an odd phenomenon. At the very moment that a crest from beam A reaches the target, the *trough* from beam B hits as well. The trough of one beam cancels the crest of the other, and vice versa. The beams interfere with each other; instead of reinforcing each other to get an extrabright beam, they cancel each other, leaving nothing behind. Instead of a bright spot on your target, you get a dark spot.

An interferometer can be a very sensitive distance-measuring device. If you stretch path A and shrink path B, you will see the spot get dark and then light and then dark again as the beams go in and out of phase. This method can tell you when the path lengths change by a fraction of the beam's wavelength, typically a few billionths of a meter. Since they

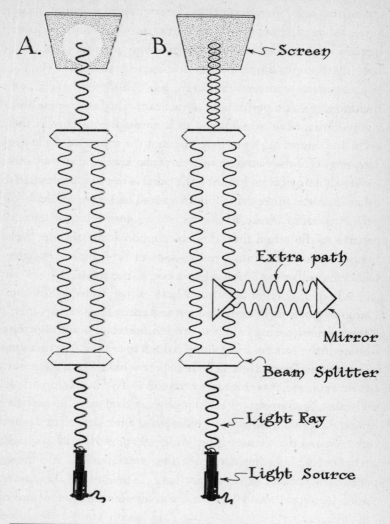

A. B.

Screen

Extra path

Mirror

Beam Splitter

Light Ray

Light Source

A light beam taking separate paths can reinforce itself or cancel itself out.

are so sensitive, interferometers are used all the time to measure distances. Surveyors, for instance, use beams of laser light to judge the size of a property; even your CD player is nothing more than a fancy interferometer.[4] The LIGO facilities are interferometers writ large. Each of the two L-shaped buildings has a powerful laser at its heart. The beam splits and zooms down both arms of the L, bounces off mirrors at the ends, and comes back to a detector near the source of the laser. The lengths of the two paths are carefully arranged so that one leg is just a tiny bit longer than the other—the two beams cancel at the detector, leaving a dark spot. The beams are locked perfectly out of phase. However, if one path changes length relative to the other, then the cancellation is messed up, and the dark spot turns into a bright spot. A brief flicker of light will signal a change in the relative size of the paths.

When a gravitational wave passes by, it squashes the L-shaped building in one direction and stretches it in the other; if the wave is coming from the proper direction, it will shorten one leg of the interferometer and stretch the other. The change in relative size sets off the telltale flicker, which is picked up by the detector. That is the theory, anyhow. The problem is that the squish-and-stretch effect of a gravitational wave is incredibly tiny, about a billionth of a billionth of a hundredth of a percent. As sensitive as laser interferometers are, they don't stand a chance of detecting a change that small, unless scientists make their yardsticks very, very long.[5] The arms of the interferometer at the LIGO facilities are about four kilometers long,

---

4. It uses a laser beam to judge the distance to the surface of the disk and whether there is a pit at a given spot or not. The pattern of pits in the disk tell the player what signals to send to the speakers.

5. Actually, there is another option, but it's not technologically feasible. If you used a light beam with an extremely short wavelength—a high-energy gamma ray for instance—your interferometer need not be quite so long, because you are measuring size changes *relative* to your wavelength. (You might notice that the Heisenberg uncertainty principle has reared its ugly head again; the higher the energy and momentum in the particles that you use to measure something, the smaller the change in size or the shorter the phenomenon you can measure.) Unfortunately, scientists can't build gamma ray lasers, so this is a nonstarter.

which means that the change in relative size will be about a billionth of a billionth of a meter. This is still too tiny a variation to
see without some major improvements to a standard interferometer. LIGO scientists have come up with a few clever mod-
ifications to make the instrument exquisitely sensitive. For
example, the light beams do not travel down the arms just
once. They bounce back and forth dozens of times before finally hitting the detector. The result is an instrument so sensitive that, despite all attempts to isolate it from its environment,
it picks up vibrations from everything around—microquakes,
tides, and even the disturbances of passing jets. This means
that noise, unwanted signals from the environment, is a
tremendous problem. However, since the two facilities are at
such a great distance from each other, most of those vibrations
can be identified as coming from nearby rather than from an
Earth-squashing gravitational wave. (A jet rattling the instrument in Washington will have no effect on the facility in
Louisiana, for example.) The facilities are almost fully operational; in 2002, scientists were busy shaking down the instruments, trying to isolate the sources of noise. Any day now, they
will be taking their first scientific data. Nobody quite knows
what LIGO will see.

Even if LIGO lives up to its design and is able to spot
squishes and stretches about a billionth of a billionth of a meter long, no one is certain that the facilities will see gravitational waves. The waves that LIGO is most likely to spot are
created by spiraling and colliding neutron stars and black
holes. Since scientists don't know how many of these pairs
are nearby, they don't know precisely how much gravitational radiation from these sorts of events is rattling around
the universe. But if and when LIGO does finally see the signature of a gravitational wave, it will be a tremendous accomplishment; it will give scientists their first direct view of
one of Einstein's spacetime ripples.

More important, gravitational radiation will become a
tool for understanding the black holes and neutron stars in

the universe. Scientists will be able to chart the heavens with gravity waves as well as light waves. However, even after LIGO upgrades its equipment around 2005—getting bulkier, more precise sapphire mirrors, for example—LIGO probably won't have the ability to see the gravitational waves that cosmologists really want to see: waves from the birth of the universe. LIGO will be an astronomer's tool, not a cosmologist's.[6]

Cosmologists want to know what happened before the era of recombination, and they realize that the process that gave rise to the structure in the universe—whether it was inflation, a big splat, or some other mechanism—was incredibly energetic. Thus, this event had to leave its mark on spacetime: gravitational waves. The universe should be chock-full of gravitational waves left over from the very earliest ages of the universe, the direct remnant of the first fractions of a second after the big bang. But since they were created very early in the life of the universe, the waves were created when the fabric of spacetime was tiny. As the fabric of spacetime expanded, these waves were stretched out to enormous size. Gravitational waves from inflation are likely to be at least tens of light-years between crest and trough, way too large for even LIGO to detect. However, big splat gravitational waves are, in theory, a bit "bluer"; they should have shorter wavelengths and might be detectable by an even more sensitive space-based version of LIGO known as LISA, the Laser Interferometer Space Antenna.

LISA will consist of three spacecraft flying in a triangular formation, more than three million miles away from one another, and will be able to detect gravitational waves compara-

6. LIGO is the most powerful of the gravitational-wave detectors. The European VIRGO and Japanese TAMA collaborations are doing the same thing, but they have shorter arms for the interferometer and are therefore less sensitive. Another set of experiments, such as ALLEGRO, which is also based in Washington state, uses large masses as tuning forks to try to pick up the passing gravitational wave of a certain frequency. This sort of experiment is less sensitive still but, in combination with LIGO, might yield some interesting data.

ble in size to the breadth of the solar system. Slated for launch in 2008, LISA may well see the leftover gravitational radiation from a big splat, and if it does, it will have seen the direct remnants of the birth of the universe.[7] However, it's possible that the gravitational ripples from the early universe are too subtle even for the most advanced gravitational-wave detector that scientists can develop in the near future. Luckily, there's another way to spot gravitational waves, and in 2002, scientists at the South Pole were taking the first steps toward spotting primordial gravitational waves. They were looking for the scars that gravitational waves leave in the cosmic microwave background.

The cosmic microwave background has appeared and reappeared in this book because cosmologists rely upon it so heavily. The ancient radiation contains an enormous amount of information about the early universe. In its hot spots and cold spots, it holds the secret to the shape of our universe, the amounts and types of dark matter and dark energy that make up the cosmos, and even the universe's ultimate fate. In addition, in late 2002 scientists began to extract another crucial bit of information about the light that scattered from the plasma during the era of recombination: its polarization.

Polarization is similar to what happens when two little girls wiggle a jump rope, making waves travel down the rope from one of them to the other. They can wiggle the rope up and down, and the ripples in the rope will wave up and down, but not left or right—the waves all lie in a vertical plane. Or, if they want, the girls can wiggle the rope left and right, and the corresponding waves will lie in the horizontal plane, rather than a vertical one. Light waves behave something like this. They can have an up-down orientation, or a left-right one, or

---

7. Unfortunately, I believe that 2008 is rather optimistic. I don't see a way for NASA to overcome the immense technical hurdles of keeping spacecraft locked in formation with the required precision, at least in the near future. I hope I'm just too pessimistic to see the solutions to such incredible hurdles.

one at any angle in between, just like the waves in a jump rope.[8] This directionality property is known as polarization, and we encounter it in everyday life. Polarizing sunglasses, for instance, exploit the polarization of light to block out half the incoming rays. A pair of polarizing glasses only allows light to pass through when a photon is oriented in the up-down direction—it is vertically polarized. If the incoming photon is horizontally polarized, it is blocked and absorbed. Liquid crystal displays also use polarization to make dark and light patches on a calculator display.[9]

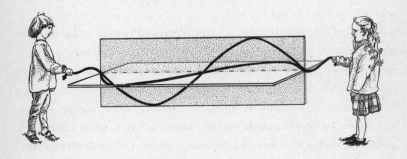

Polarized jump ropes

Polarizing sunglasses are popular because they block road glare. Glare occurs because light becomes polarized when it bounces off a surface. Ordinarily, all the photons in a light beam have a random polarization. Some are polarized horizontally, some vertically, and the rest are in between. As a whole, the beam does not have a preferred polarization. However, when you are driving along a (horizontal) road, the sun's rays strike the road and pick up a polarization parallel to the surface

8. The girls can wiggle the jump rope in other ways too, such as in a circle, but such motions are actually a superposition of left-right and up-down wiggling in different proportions and at different times. We need only talk about the *linear* polarization without worrying about the *circular* or *elliptical* types.
9. You can see this if you have a calculator and a pair of polarizing sunglasses. Look at the display through the glasses and then rotate the calculator with respect to the glasses. You will see the display fade in and out as their polarizing filters alternately become aligned and misaligned.

they bounce off of—the beams become horizontally polarized. So the sun's glare, all the stuff that reflects off the road, tends to be horizontally polarized, which is precisely what the polarizing lenses block. Glare doesn't stand a chance when it hits a polarizing filter.

In the moments before recombination, when the electrons finally combined with nuclei, photons were constantly bouncing about, scattering off the particles in the plasma, skittering to and fro, unable to keep moving in one direction for very long. When the plasma cooled and the electrons recombined, each photon took its last bounce. However, like any photon bouncing off an object, it picked up a polarization when it scattered that final time. The polarization of these photons—the light that would become the cosmic background radiation—was parallel to the *last scattering surface,* the cloud of plasma that forms the fiery walls in the heavens. The polarization encodes information about the universe at recombination, just as the temperature fluctuations do. In fact, polarization is better than temperature fluctuations for revealing the state of the early universe.

The cosmic background radiation that scientists observe has been on a fourteen-billion-year journey, a tough journey indeed. As the radiation travels, it is stretched and kneaded. When a photon approaches a cluster of galaxies, it sinks into the dimple in spacetime caused by the cluster's enormous mass, stretching out and getting slightly bluer as it dips deeper and deeper into the dimple. As it rises out of the dimple, it gets redder again. All in all, the effects upon descending into and ascending from the dimple should cancel. However, the dimple is also changing shape and size over time, so the effects don't quite cancel out, leaving a residual stretch or squash to the photon, raising or lowering its temperature a tiny bit. The photons from the last scattering surface undergo a tremendous amount of gravitational kneading as they travel toward Earth. This kneading, known as the Sachs-Wolfe effect, leaves its imprint on the cosmic background radiation, changing the photons' temperatures slightly and contaminating the signal somewhat.

Though this contamination can yield some very useful information about the development of the early universe, it makes the signal from the last scattering surface less pure than it would otherwise be, diluting the information from the early universe. However, a photon's polarization, unlike its temperature, is unaffected by dimples in spacetime. Even as the photons get hotter and colder, the polarization stays the same. By studying the angles of the polarization of the cosmic microwave background, scientists can calculate where the light was bouncing from, information that reveals how matter was distributed in the early universe—a signal that is much more precise than the contaminated temperature measurements. "Polarization is a lot cleaner," says John Kovac, a physicist at the University of Chicago. "It directly probes the last scattering surface."

But scientists first had to detect the polarization before they could use it, and measuring temperature fluctuations of a few millionths of a degree is quite simple by comparison. Nevertheless, in September 2002, Kovac and his colleagues succeeded. Careful observations with the DASI telescope at the South Pole revealed the polarization of the cosmic background radiation for the first time. "It's like going from the picture on a black-and-white TV to color," says Kovac.

Polarization gives cosmologists another measurement of the amount of stuff—all the matter and energy—that populates the universe, its distribution, and how it was moving. That is not the only result, though. Polarization will also give astrophysicists a tool to probe another era in the early universe: reionization, the time a few hundred million years after the big bang when enough stars, galaxies, and quasars ignited and burned off the hydrogen fog that caused the cosmic dark age. This effect should appear as a small peak in the cosmic microwave background spectrum on large angular scales. Gravitational kneading erases the peak from temperature-variation data, but polarized light should make it visible, says Max Tegmark of the University of Pennsylvania. The information will reveal how long ago reionization happened. "It's one of the most exciting numbers that we have no clue about right now, and there's no way we can do that with current measurements," Tegmark explains.

Yet there is a still more exciting signal hiding within the polarization of the cosmic microwave background. The radiation is scarred by gravity waves, the ripples in spacetime that hearken back to the first fractions of a second after the birth of the universe. Though a photon's polarization is not affected when it dips in and out of a dimple in spacetime, the peculiar squash-and-stretch action of a gravitational wave does affect its polarization. The signature of a gravitational wave from inflation (or from the big splat) in the cosmic microwave background is a spiral quality that mathematicians call curl. In a hypothetical map of the polarization of the cosmic background radiation, this curl-type component would look something like little hurricanes. Mere acoustic oscillations cannot produce any of these spirals, but inflationary theory predicts that gravitational waves in the early universe *must* have created a curl-type component in the cosmic background radiation. If scientists studying light from the cosmic microwave background see a clean curl-containing component, one uncontaminated with false signals from, say, polarized light from galaxies, it will be "a smoking gun for gravitational waves," says Tegmark.

Unfortunately, the spirals in the sky are extremely faint, so astronomers don't expect to see them for several years. However, in 2007, the European Space Agency is launching a microwave telescope named Planck. A successor to the Microwave Anisotropy Probe, a satellite that released the first precision all-sky map of the CMB as this book went to press in February 2003, Planck will have a sensitive polarization detector. It should spot the spirals in the sky—the signature of gravitational waves from the very beginning of the universe. "We'll be probing the universe at 10^{-30} of a second," says DASI's John Carlstrom. Such an observation could make or break inflation theory. "Gravitational waves might tell whether inflation is right, or if something else, like the big splat is," says Princeton's P. J. E. Peebles. "We can certainly hope so." Within a decade, we may see the very face of creation.

Chapter 14
Beyond the Third Revolution

[VOYAGE TO THE ENDS OF TIME]

And there, there overhead, there, there hung over

Those thousands of white faces, those dazed eyes,

There in the starless dark the poise, the hover,

There with vast wings across the canceled skies,

There in the sudden blackness the black pall

Of nothing, nothing, nothing—nothing at all.

— ARCHIBALD MACLEISH, "THE END
OF THE WORLD"

For the first time, scientists are answering the questions that have plagued humanity for millennia. How did the universe begin? How will it end? Astrophysicists are beginning to illuminate the first moments of creation, and barring the victory of the ekpyrotic scenario over the big bang theory, they know how the cosmos will die. However, these triumphs will not mark the end of cosmology. There are still places to seek answers.

Some physicists are trying to figure out the mechanism of

the big bang itself, trying to understand what laws of physics gave birth to our cosmos. Yet others are looking far into the future and asking whether civilization can survive indefinitely in a decaying and expanding universe, or whether life itself is doomed. Still others are trying to divine whether our own is the only universe, or whether an infinite number of universes are out there, each with its own properties. And some are looking for the hand of the creator.

These questions are currently well beyond the realm of experiment; nothing that scientists can conceive of can test the exotic theories proposed to answer them, yet farsighted physicists are researching them today. They are the stuff of the next revolution. These are the questions that will take us to the edges of space and time.

To understand the forces that gave birth to the universe, physicists have to patch the holes in their understanding of the forces of nature. One of the fundamental problems in physics is the conflict between quantum theory—the equations that govern very small things like electrons and protons—and general relativity, the theory that rules very large and massive objects like stars and galaxies. The two theories are incompatible, and where their domains intersect, relativity and quantum mechanics clash—for instance, at a black hole. It is a huge amount of mass, which is best dealt with through the equations of relativity, but the mass is compressed into a very small space, which catapults it into the realm of quantum mechanics. As a result, nobody knows precisely what happens at the very heart of a black hole. The primordial universe of the big bang, like the black hole, is a mystery beyond the reach of science's physical models. During the big bang, the entire universe and all of its matter and energy had to grow from a tiny subatomic seed. Scientists simply don't know what equations they should use to describe such a tiny, dense object; they don't know what laws of physics held during the birth of the universe.

The clash between quantum mechanics and relativity is caused, in part, by their very natures. Relativity is a theory that deals with smooth surfaces, like rubber sheets. In quantum mechanics, however, nothing is smooth. Objects move in discrete jumps; they are quantized rather than continuous. What happens when you look at the sheet of spacetime on a very small scale? Is it smooth, as Einstein says, or is it choppy, as quantum theory would imply? No one is sure. All that scientists know is that at very small scales, and at very large energies, quantum theory and relativity stop working. The laws of physics break down.

Einstein devoted the latter half of his life to reconciling quantum mechanics with his theory of relativity. He hoped to find a grand unified theory that would explain all phenomena, at all scales, without the contradictions that present theories all have. He failed. When Einstein died, science was no closer to a theory of quantum gravity than when he started his quest for the theory of everything. However, there is hope on the horizon. Over the past few decades, theoretical physicists like Edward Witten of Princeton and Juan Maldacena of Harvard have been working on a theory of quantum gravity, and it seems to resolve the conflicts between relativity and quantum mechanics. Instead of treating particles, like electrons, as points in four-dimensional spacetime, the new theory, M-theory, considers these particles as membranes in eleven-dimensional space.[1] As nonsensical as this theory seems to a nonphysicist, it is gaining adherents by the day, because it solves some major problems in physics. Not only does it resolve the conflict between quantum theory and relativity, but it also incorporates the theory of supersymmetry and unifies the strong, electroweak, and gravitational forces. It is mathe-

1. Most of these dimensions are *compactified* or curled up so we don't perceive them; they don't really "mean" anything in the way that the dimensions we're familiar with do. There are a few interesting variations on theories out there, notably those proposed by Nima Arkani-Hamed at Harvard and Andreas Albrecht of the University of California at Davis. In these theories, some of those extra dimensions are relatively large—millimeter size, even—which would have some observable consequences.

matically beautiful, and more important to cosmologists, it gives scientists the tools to understand the physics at the very heart of a black hole, and of the big bang. If M-theory is correct, scientists might finally unravel the mystery of the big bang itself. They will be able to use their equations to explore it directly, something they are unable to do with today's physical laws. Nevertheless, even if M-theory is correct, they may never be able to test it.

The problem is that these membranes and these extra dimensions are very, very small. Thanks to the Heisenberg uncertainty principle, a huge amount of energy is required to probe objects of such small size, so scientists would need an enormous accelerator to verify M-theory directly. How enormous? With today's magnet technology, the particle accelerator would have to be six million billion miles around. Even traveling at the speed of light, a particle would take one thousand years to complete a circuit. This is not an option. The best hope for confirming M-theory lies with the cosmic background radiation. If scientists are very lucky, they might be able to confirm the big splat scenario or some variant thereof; since the big splat uses the laws of M-theory as its basis, a confirmation of the big splat theory's predictions would be a compelling piece of evidence that M-theory is correct. Such a confirmation, however, is at least a decade away, and would require quite a bit of luck—and funding. It is also possible that a confirmation of M-theory will be well out of reach for many more years. Scientists might have the right answer and never know whether it's a fiction or reality. After all, we have a finite time in the universe to test these theories.

How finite? In about a billion years, the sun, which slowly heats up as it burns its fuel, will warm the Earth, evaporating the oceans in a runaway greenhouse effect. Earth will turn into another Venus, sweltering and lifeless. It is possible that by then civilization will have mastered long-distance spaceflight. Humanity might scatter itself among the stars of the galaxy. However, even those stars have a finite lifetime. Now that we think we know the end of the universe—that it will expand forever, cooling and dying—we are forced to confront our ultimate fate. Can a civilization survive indefinitely in a dying universe, or must life be snuffed out in the ever-cooling soup of lifeless particles? Physicists are trying to find out.

The far-future universe would be a bleak place indeed. As the universe expands ever faster, distant galaxies will redden and dim. They will disappear from view, the most distant first. Soon, even nearby galaxy clusters will disappear. The night sky will become emptier and emptier.[2] As stars in each galaxy burn out and die, the galaxies we can see become darker and darker. Energy becomes harder to find.

Life, and consciousness itself, runs on energy. The laws of physics state that even a nonliving organism that can calculate, that can do computations, must expend energy as it computes.[3] As energy gets harder to find, any civilization that hopes to survive must go on an energy "diet," but that diet restricts how much thinking that beings in the civilization can do. It seems hopeless: a civilization slows down more and more, thinking less and less, until it ceases entirely.[4] In 1979, however, physicist Freeman Dyson proposed a way to keep a civilization running, even as its energy supplies burn out: hibernation. A civilization might alternate periods of activity with ever increasing periods of hibernation. During the periods of hibernation, the civilization's machines would gather and store energy. When enough energy had been collected, the beings would wake up, use that energy to live, and when their supplies ran low, drop back into hibernation. Even though the periods of hibernation would get larger and larger, eventually stretching into millions and billions of years, the inhabitants of such a distant-future civilization might be able to

2. Since there will be less and less to observe in the heavens as time goes on, Michael Turner of the University of Chicago urges, "Fund cosmology now!"

3. This is so long as it has a finite memory. Oddly enough, the expenditure of energy comes when a computing device *erases* a previously used spot in memory. (This strange law was discovered by the late Rolf Landauer of IBM.)

4. Some would say the process has already started.

survive indefinitely. However, it now seems that not even Dyson's scheme can save civilization from dying.

Lawrence Krauss, a physicist at Case Western Reserve University, recently showed that even with such a hibernation scheme, a civilization cannot carry on forever. Even as the hibernation periods get longer and longer, the corresponding periods of activity have to get shorter and shorter. These periods of activity shorten so dramatically that the civilization functionally has a finite lifetime. After a certain date, all the energy available for the rest of the bleak but infinite life of the universe cannot keep the civilization going for even a second more. There are a finite number of thoughts that civilizations can have. As energy runs out, they must stop thinking and die. Life cannot be eternal in our universe.

But even this might not be the end of life. Other scientists hope that life might go on if there are other universes as well as our own. This is not as far-fetched as it might seem. Inflation theory says that we live in an expanding bubble of true vacuum; there are probably other bubbles out there, unobservable and separated from our own universe by a wall of false vacuum. They would hardly count as entirely separate universes; these bubbles were created by the same big bang as our own, and even though we are unable to communicate with them at the moment, in theory these bubbles might eventually merge as they expand.

Some scientists envision a more extreme scenario that makes mere bubble universes seem insignificant. These theorists believe that the laws of quantum mechanics are spawning an uncountable number of entirely new universes at every moment. This strange idea has its roots in one of the most counterintuitive aspects of quantum mechanics: the principle of superposition. In the ordinary world of classical physics, an object cannot take two *states* at once; a switch has to be either up or down, a top has to spin clockwise or counterclockwise, and a cat has to be either alive or dead. In the quantum world, though, this isn't true. A photon can go through a left slit *and* a right slit in a barrier at the same time, an electron can have spin up and spin down, and Schrödinger's cat, if sufficiently isolated from his environment, can be both alive and dead.[5] However, once information about the object begins to seep into the world, say, through an observer looking at a cat or measuring the electron, the object must "choose" which state it is in. You will never see an alive-dead cat; when you open the box, it will be either alive or dead, not both. But such a strange situation leads to some troubling properties of quantum objects.

This property of quantum objects has led some scientists, like David Deutsch at Oxford University and Sir Martin Rees, the Astronomer Royal, to investigate whether some of the paradoxes of quantum mechanics can be solved by a peculiar assumption. If our universe is part of an enormous *multiverse* that constantly proliferates and sprouts new universes, the rules of quantum mechanics begin to make a bit more sense. One form of this *many worlds* hypothesis states that every time a quantum object makes a "choice"—alive or dead, spin up or down, left slit or right slit—our universe splits in two. In one universe, Schrödinger's cat lives, and in the other it dies. Though this scenario seems needlessly complicated, it is a mathematically valid interpretation of quantum mechanics, and it provides some answers to the fine-tuning question: why is our universe so ideally suited for life?

A number of constants in the universe govern the way matter and energy behave. The speed of light, for instance, governs how fast things can move across the surface of space-time. The gravitational constant dictates how strong the force of gravity is. There is a handful of these fundamental con-

5. At least for a tiny fraction of a second. Scientists are looking at what makes a quantum object "choose" and end its superposition. This subject, decoherence, is a very hot topic in quantum mechanics. The larger something is, and the less isolated it is from its environment, the quicker it decoheres and loses its superposition. A large, warm, sloppy cat is unlikely to hold its superposition for very long, even in the absence of an observer.

stants, and if any one of them were significantly different, life probably could not have happened.[6] If gravity were too strong, suns would be incredibly massive, if they existed at all, burning brightly for a very short time. Too weak and they would seldom ignite, forming galaxies full of brown dwarfs. We live in the happy medium. Too happy, for some.

It seems like a tremendous coincidence that the universe is suitable for life. If you were to choose values for these constants at random, life probably would not exist. It is a stunning coincidence that the universe is the way it is. Scientists tend to be uncomfortable with coincidences, and the many worlds interpretation gives a way out. If the many worlds scenario is true, there might be lots and lots of different universes out there with different constants. Some collapse in a millisecond. Some have almost no matter in them. We just happen to inhabit one that is suitable for life.[7]

Others believe that the coincidence is just that—a coincidence.[8] Still others believe that the fine-tuned universe is the signature of a creator. The John Templeton Foundation, an organization devoted to exploring the "spiritual dimensions" of the universe, granted a million dollars to researchers ex-

6. In 2001, scientists in Australia and the United States saw evidence that one of these constants, the fine-structure constant, which is related to the strength of electromagnetic interactions, might have changed very slightly over billions of years. In 2002, another team of scientists showed that a changing fine-structure constant means that the speed of light was probably changing as well. Though these observations are taken seriously by the astrophysical community, the consensus is that there are subtle problems with the observations—the fine-structure constant probably is not changing after all. Nevertheless, it is something that physicists will keep an eye out for in the future.

7. In fact, the probability of winding up in one suitable for life is pretty great, because those universes tend to be the most complex—they have the most choices. Each choice is like a branch on a tree; some universes are tiny stunted saplings and others are towering pines with an unimaginable number of branches. The simple universes, the ones that collapse or have very little in them, don't have very many branches. If you look at sheer numbers (ignoring the fact that we're dealing with infinities here), any branch we sit on is likely to be from a complex universe, which is much more likely to harbor life than a simple one.

8. This coincidence is not as troubling as it looks, because our very existence, and our ability to wonder at the state of the universe, is predicated upon the universe's being habitable. Unfortunately, this argument, known as the weak anthropic principle, yields no new information, though it does rob the fine-tuning coincidence of its sting.

ploring scientific matters related to fine-tuning. The Templeton Foundation is trying to find God in the cosmology revolution—and this is where the science ends and philosophy
begins.

Right now, these subjects are in the realm of philosophy
and religion, beyond the reach of experimental science. But
just as the puzzles of ancient cosmology, the Greek and Christian ideas about the universe, moved from the realm of philosophy to the testable world of science, these issues too might
be answered by a future generation of scientists. They are
subjects, perhaps, of a fourth cosmological revolution.

The third is still far from over, though scientists have already answered one of the biggest questions that have plagued
humans since the dawn of civilization. We now know roughly
what the universe is made of, and we know how it will end.
This stunning achievement was not an ending, but a beginning. By 2010, the end of the decade, when the current cosmological revolution is complete, we will have detailed
answers about where the universe came from and where it is
going. Physicists will have seen the presence of dark matter,
both ordinary and exotic. They will begin to understand the
mysteries of the vacuum and the physics of dark energy and
the early inflation of the universe. They will have released
quarks from their colorful confinement and seen why our universe is made of matter rather than antimatter.

A decade from now, we will look back and realize how
much our view of the universe changed when scientists gazed
into the face of creation. We will understand the beginning
and the end—the alpha and the omega.

Appendix **A**
Tired Light Retired

There is little way to deny the big bang theory if you accept the fact that the reddening of galaxies is caused by the Doppler effect. Galaxies are speeding away, the more distant the faster, so the universe is expanding. But a small band of contrarians argue that the reddening is not caused by the Doppler effect. They argue that galaxies' light reddens because it loses energy as it passes through space: light gets "tired." This tired-light hypothesis was invented by astrophysicist Fritz Zwicky within a few months of the publication of Hubble's paper on the expansion of the universe. Zwicky wanted to explain the reddening of distant galaxies without resorting to an ever expanding universe. In his tired-light scenario, distant galaxies are red not because they are moving, but because their light has traveled farther and gotten pooped out along the way.

When experimenters first measured the cosmic microwave background in the 1960s, they found that the radiation was too dim to be explained by Zwicky's hypothesis. That realization relegated tired light firmly to the fringe of physics, but scientists still sought more direct proofs of the expansion of the cosmos. Two papers released in 2001 provide the best direct evidence yet.

The first measures the brightening and dimming of supernovae. Thanks to Einstein's theory of relativity, we know that if distant supernovae are moving away at great speeds, their "clocks" tick more slowly than one on Earth, because of the phenomenon of time dilation. As a result, distant supernovae seem to explode and evolve in slow motion—they will appear to flare and fade at a more leisurely pace than nearby ones. A team of scientists led by Gerson Goldhaber of the Lawrence Berkeley National Laboratory (LBNL) in Berkeley, California, has shown that this is, indeed, the case with forty-two recently analyzed supernovae. "It's very unambiguous," says LBNL supernova hunter Saul Perlmutter.

In the second study, Allan Sandage of the Carnegie Observatories in Pasadena and Lori Lubin, currently at the University of California at Davis, analyzed space-based measurements of the surface brightness of galaxies. Both the standard expanding-universe theory and the tired-light theory, they realized, predict that redshifted light should make distant galaxies look dimmer than they really are; since redder light is less energetic, galaxies will look dimmer no matter whether the reddening comes from tired light or from the motion of the galaxies. However, a galaxy will appear *much* dimmer at great distances if it is moving, for two reasons that do not apply to stationary galaxies.

The first reason, as in the supernova paper, is relativistic time dilation. Imagine that a galaxy spits out a photon toward Earth every second. Because the clock of a distant, moving galaxy is slow compared with one on Earth, the photons are more than a second apart from Earth's point of view; fewer photons arrive in any given span of time, so the galaxy appears dimmer. The second reason is a phenomenon known as relativistic aberration, which distorts the apparent shape of the galaxy, making it appear much dimmer than it would be if it were stationary. These two effects, time dilation and aberration, only apply to moving, Doppler-shifted galaxies, not to stationary, tired-light ones.

Sure enough, when Sandage and Lubin measured the surface brightness of a number of galaxies, they found that the galaxies were much dimmer than the tired-light theory would suggest, and taking into account the fact that distant galaxies are a bit brighter than those nearby (because the ancient galaxies were populated by bright, young stars), the observation matched the moving-galaxy brightness prediction quite nicely.

"The expansion is real. It's not due to an unknown physical process. That is the conclusion," says Sandage. Tired-light theory has been thoroughly retired. Hubble was right: the universe is expanding.

Appendix **B**
Where Does Matter Come From?

Symmetry and asymmetry are powerful tools in the hands of the particle physicist. Indeed, the whole structure of the subatomic world seems to be built upon symmetries, and perhaps even supersymmetry, as discussed in chapter 10. The discovery of a new symmetry in the universe, or of the breakdown of a seemingly established symmetry, is usually the signal of a new fundamental truth about the way the cosmos works. Three of the key symmetries of particle physics are known by their initials: C, P, and T. These three symmetries seem to hold the secret of the difference between matter and antimatter.

When Alice took a trip through the looking glass, she entered a world where everything was reversed, as if it were reflected in the mirror. When she saw the text of the poem "Jabberwocky." the letters and words went from right to left instead of from left to right. She had gone through a mirror reflection. The essence of P symmetry (the P stands for "parity") is that Alice wouldn't notice a difference in the laws of physics of her home world and the looking-glass world; after reflecting the universe in a mirror, the laws of physics stay the same.[1]

1. Parity is a mathematical term that has to do with symmetries in space. Technically, P symmetry is a reflection in three mirrors, rather than just one. You switch left for right, up for down, and front for back.

When P symmetry fails, the laws of physics are slightly different in the looking-glass world.

Until the late 1950s, scientists thought that P symmetry was a fundamental rule that governed the universe; if you somehow magically reflected the universe in a supermirror, the two universes would always be indistinguishable. However, on the subatomic scale, mirror-reflected matter has subtle differences from ordinary matter. Chen Ning Yang and Tsung-Dao Lee, working at the Institute for Advanced Study in Princeton, New Jersey, proposed a way to test whether P symmetry was violated in certain nuclear decays. Their experiment (which Chien-Shung Wu of Columbia University carried out) set up a situation where decaying cobalt nuclei spit out electrons both upward and downward. The result: more electrons traveled downward than upward. They also showed that if they had done the experiment in the mirror world, there would have been more upward-traveling electrons than downward-traveling ones. So their experiment revealed a difference between our universe and a mirror-image one; in one universe, more electrons traveled up, and in another, they traveled down. P symmetry was violated, because the mirror universe was not identical to our own. For this, Yang and Lee received the 1957 Nobel Prize in Physics.

For a time, physicists thought they could reinstate P symmetry by adding another condition, called C symmetry. Just as P symmetry has to do with replacing matter with mirror matter, C symmetry (the C stands for "charge") replaces matter with antimatter. The combination of the two symmetries, CP symmetry, asserts that the laws of physics would remain the same if you replaced matter with antimatter *and* reflected the universe in a mirror. (The third type of symmetry, T, which stands for "time," involves hypothetically doing an experiment in reverse.)

In Yang and Lee's experiment, CP symmetry was true. If you replaced matter with antimatter and mirror-reflected the setup, the results would remain the same. Yang and Lee's experiment showed a violation of P symmetry, but CP symmetry held fast—for another few years. In 1964, Val Fitch and

James Cronin published a paper in *Physical Review Letters* that showed that the K^0 meson (made of a down quark and a strange antiquark) decayed in a way that can only happen if CP symmetry is violated. The failure of CP symmetry holds the secret to understanding where matter comes from.[2]

A violation of CP symmetry can manifest itself in many different ways. Recent experiments at Fermilab in Batavia, Illinois, studied the decay of K mesons; in particular, the researchers paid careful attention to the angles at which the decay products flew away. Violations in CP symmetry show up as preferred angles, just as P-symmetry violations show up as electrons' preference to fly down rather than up. But another manifestation, measured at CERN, is even more striking. In May 2001, after a decade of work, a collaboration at CERN presented measurements of twenty million K^0 meson and anti-K^0 meson decays. Anti-K^0 particles decayed just a tiny bit faster than K^0 particles. This means that if you could create a bunch of K mesons and anti-K mesons, you would see the antimatter versions wink out of existence before the matter versions do. In 1967, Russian physicist Andrei Sakharov proposed that this tiny asymmetry gives matter a slight, but crucial, edge over antimatter.

When the universe was born, presumably the energy of the big bang went into creating matter and antimatter in roughly equal proportions. If the amount of matter had been truly equal to the amount of antimatter, then the matter and antimatter in the universe should have annihilated each other, leaving nothing behind but a soup of energy. But since matter seems to have a small edge—matter lasts a little bit longer than antimatter and is thus "preferred" by nature—a tad more matter than antimatter survived, perhaps one part in a

2. Scientists now think that the symmetry that holds true is CPT symmetry, where the T symmetry indicates that if the flow of time is reversed in a doppelgänger universe in addition to matter being switched with antimatter and reflected in a mirror, the new universe will be indistinguishable from our own. Nobody has yet found any signs of a violation of CPT symmetry.

billion. That extra little bit is our inheritance. As antimatter and matter annihilated each other, the extra little bit remained and became the matter that forms the universe.

Scientists don't have a full handle on the process of CP violation yet. For a long time, the K meson was the only particle that showed CP-violating tendencies, and to get a full mathematical portrait of the CP-violation process, it must be observed in another type of particle that contains a more exotic quark, like a bottom quark.[3] Since bottoms are massive quarks, they are rare, and the particles that contain bottom quarks (and bottom antiquarks), like the B meson, are hard to make.

However, it is not impossible. For the past few years, the Stanford Linear Accelerator Center in California has been generating swarms of Bs and watching them decay. Another B factory in Japan has been doing the same thing. In 2001, the first results began to pour out. Sure enough, the teams in both California and Japan saw hints of CP violation in B mesons, and pretty soon the Tevatron accelerator at Fermilab will be producing scads of Bs as well. It is still too early to make a definitive statement, but scientists are almost ready to announce that they have finally found the last piece in the CP-violation puzzle. With the observation of CP-violation in B mesons, scientists will be able to paint the full mathematical portrait of the CP-violation process in quarks—and how the universe came to be populated with matter instead of antimatter.

3. The mathematics of CP violation relies upon an object known as the Cabibbo-Kobayashi-Maskawa (CKM) matrix. This matrix encodes certain types of interactions between the quarks, and the terms in that matrix aren't fully known, particularly the ones that involve CP violation. K mesons reveal some of the terms in the matrix, but another term is needed to figure out the entire matrix. This is why scientists need another particle to get a complete picture of the CP-violation process.

Appendix **C**
Nobel Prizes in Physics—
Past and Future

The Nobel Prizes referred to in the text are as follows:

1933: P. A. M. Dirac, for the prediction of the antielectron. (Also Erwin Schrödinger, for quantum mechanics.)

1936: Carl Anderson, for the discovery of the antielectron. (Also Victor Hess, for the discovery of cosmic rays.)

1957: Chen Ning Yang and Tsung-Dao Lee, for the discovery of P violation in cobalt decays.

1965: Sin-Itiro Tomonaga, Julian Schwinger, and Richard Feynman, for renormalization in quantum electrodynamics.

1969: Murray Gell-Mann, for quantum chromodynamics.

1974: Antony Hewish, for discovering the first pulsar. (Also Martin Ryle, for inventing synthetic apertures—a technique related to interferometry.)

1976: Burton Richter and Samuel Ting, for discovering the J/psi meson.

1978: Arno Penzias and Robert Wilson, for the discovery of the cosmic background radiation. (Also Pyotr Kapitza, for low-temperature experiments.)

1979: Sheldon Glashow, Abdus Salam, and Steven Weinberg, for electroweak unification.

1980: James Cronin and Val Fitch, for the discovery of CP violation in K mesons.

1984: Carlo Rubbia and Simon van der Meer, for the detection of W and Z bosons.

1988: Leon Lederman, Melvin Schwartz, and Jack Steinberger, for the detection of the muon neutrino.

1993: Joseph Taylor and Russell Hulse, for the discovery of a binary pulsar, which confirmed the existence of gravitational waves predicted by Einstein's theory of relativity.

1999: Gerardus 't Hooft and Martinus Veltman, for renormalization of electroweak theory.

2002: Raymond Davis Jr. and Masatoshi Koshiba, for the detection of solar and cosmic neutrinos. (Also Ricardo Giacconi, for pioneering work in x-ray astronomy.)

It is always difficult to guess the intent of the Nobel committee, and harder still to figure out who, within a large field, is going to get a prize for a particular discovery. The committee often lets political or philosophical bias get in the way of giving the awards to deserving candidates. Edwin Hubble never won a Nobel Prize, and Einstein's prize, for an explanation of the photoelectric effect, was given *despite* his theory of relativity. Only two things are certain: at most three people can share a prize, and nobody can be awarded one posthumously.

Nevertheless, the past few years have seen a great deal of cosmology-related work that is worthy of Nobel Prizes. Here are my predictions about completed research that will eventually be awarded a Nobel, and my best guesses as to who will take home the prize:

For the discovery of dark matter (Vera Rubin and others)

For inflationary theory (Alan Guth)

For the discovery of anisotropies in the cosmic background radiation (Members of the COBE team, and possibly others)

For the accurate prediction of the power spectrum of the cosmic background radiation (Edward Harrison, P. J. E. Peebles, J. T. Yu, or others—Yacov Zel'dovich died in 1987)

For precision measurements of the power spectrum of the cosmic background radiation (Members of the Boomerang and DASI teams)

For the discovery of neutrino mass (Members of the Super-Kamiokande team)

For prediction of the spectrum of solar neutrinos from the sun (John Bahcall and others)

For the solution of the solar neutrino paradox (Members of the Sudbury Neutrino Observatory and the Super-Kamiokande teams)

For the discovery of dark energy (Members of the High-Z Supernova Search Team and the Supernova Cosmology Project)

For measuring the curvature of the universe (Members of the High-Z Supernova Search Team, the Supernova Cosmology Project, and Boomerang)

More difficult still is predicting Nobel Prizes for work that is yet to be completed, although the third cosmological revolution is pregnant with possibilities, including the following:

For the prediction and discovery of supersymmetric particles

For the creation and analysis of a quark-gluon plasma

For the prediction and discovery of curl-type polarization in the cosmic background radiation

For the identification of dark matter objects in the halo of the Milky Way

For the discovery of a new weakly interacting massive particle that contributes significantly to dark matter

For the discovery of the Higgs boson

For the analysis of weak decays in B mesons and the completion of the CKM matrix

For the discovery of double-beta decay and the proof that the Majorana picture of the neutrino is correct (unlikely, but if found, a certain Nobel)

For the direct detection of gravitational waves

Appendix D
Some Experiments to Watch

A number of exciting experiments were ongoing in 2002 in the five fields listed below. This is merely a selection, and a taste of things to come.

Cosmic Microwave Background
(Cosmic Background Radiation)

Boomerang: This Antarctic balloon-based observatory has already revolutionized the field of CMB astronomy. First deployed in early 1999, Boomerang (the name is derived from Balloon Observations of Millimetric Extragalactic Radiation and Geophysics) provided the first highly detailed measurements of the background radiation. It has undergone a refit to make it sensitive to polarization and will likely return results shortly.

DASI: Like Boomerang, DASI (the Degree Angular Scale Interferometer) is a sensitive Antarctic observatory. However, it is ground-based and uses interferometry rather than bolometers to do its measurements. DASI scientists first released high-quality data in April 2001, and in September 2002 DASI was the first instrument to detect the polarization of the cosmic microwave background.

CBI: The Cosmic Background Imager is similar to DASI, but it is based in Chile and is sensitive to smaller angular scales compared with DASI and Boomerang. While less celebrated than DASI and Boomerang, CBI has already provided significant support for inflationary theory and is likely to make important observations over the next few years that DASI and Boomerang are unable to provide.

MAP: Launched aboard a Delta II rocket in June 2001, the Microwave Anisotropy Probe has been taking high-resolution pictures of the cosmic background radiation over the entire sky, unlike Boomerang, DASI, or any other earthbound telescope, which can only take a picture of a section of the sky. Such a detailed and comprehensive map will pin down the spectrum of the CMB with unprecedented precision. The first results arrived in February 2003, as this book was going to press.

ACBAR: First deployed at the South Pole in November 2001, the Arcminute Cosmology Bolometer Array Receiver is intended to take advantage of the Sunyaev-Zel'dovich effect to map out the distribution of matter in galaxy clusters. In 2006, an as-yet-unnamed telescope at the South Pole will perform a much more comprehensive Suryaev-Zel'dovich survey of the skies.

Planck: Scheduled for launch in 2007, this European satellite, like MAP, will observe the cosmic background radiation across the whole sky. Not only will it be more precise than MAP, but it will also be able to detect polarization, which MAP cannot do with any real resolution.

Astronomical Observations

2dF: The Two Degree Field collaboration is using an Australian telescope to map galaxies and other objects in the sky. The project's astronomers expect to map 250,000 galaxies, and they have almost reached their goal. The 2dF's data have already revealed the distribution of matter in galaxy clusters, and the data are expected to get even better.

SDSS: The Sloan Digital Sky Survey is very similar to 2dF in its goals and methods; however, it is a more extensive search. SDSS researchers have already released some valuable data and are expected to return a lot more over the next few years.

SNAP: The Supernova Acceleration Probe is a proposed satellite that would use a very high-tech camera to take snapshots of the heavens in hopes of spotting supernovae, particularly type Ia supernovae. If it is launched, it would immediately reward the supernova hunters with a rich bounty of standard candles and allow cosmologists to figure out the rate of the expansion of the universe over an enormous time period.

High-Energy Physics/Particle Physics

RHIC: The Relativistic Heavy Ion Collider at Brookhaven National Laboratory smashes heavy nuclei, like gold, into each other at enormous speeds. Since the machine began operations in June 2000, evidence has indicated that the machine has produced a quark-gluon plasma, though the data are not strong enough for RHIC scientists to make a definitive claim. Expect an announcement of the discovery of a quark-gluon plasma in 2004.

BaBar: Based at the Stanford Linear Accelerator Center in California, BaBar is an instrument that analyzes B mesons.

The first data arrived in 1999 and results have been trickling out since then. These measurements will help scientists fill in details about weak interactions and CP violation and will help explain why our universe is made of matter rather than antimatter.

Tevatron: After a $260 million refit, the Tevatron accelerator at Fermilab is having some teething troubles. Since it was turned on in March 2001, the proton-antiproton smasher has not been performing well. Once the kinks are worked out, however, the Tevatron is likely to pin down some of the details of the W bosons and should produce numerous B mesons, adding to the knowledge provided by BaBar. Furthermore, there is a very good chance that Tevatron will spot the lightest supersymmetric particle, and a slim chance that it will see the Higgs boson.

LHC: The Large Hadron Collider, at CERN in Geneva, Switzerland, will exceed the capabilities of Tevatron and RHIC. If the lightest supersymmetric partner is not found by the Tevatron, then it will be found by the LHC, or supersymmetry will be all but ruled out. It should also find the Higgs boson. The LHC is scheduled to be operational in 2007 but will likely be delayed.

NLC: The Next Linear Collider is a proposed $6 billion facility meant to complement the LHC; if approved, it probably will be built on the West Coast of the United States or in Germany. Unlike the other accelerators described here, the NLC would smash positrons and electrons together, instead of composite particles like protons or nuclei. This makes the NLC a scalpel to the LHC's chainsaw. Once the LHC spots a particle of interest, the NLC would be able to investigate its properties in great detail. Such an expensive project is likely to face a rocky road ahead, but if it is commissioned, it will be a spectacular instrument.

Gravitational Waves

LIGO: The Laser Interferometer Gravitational-Wave Observatory is a set of two facilities designed to detect the distinctive stretch-and-squash signature of a passing gravitational wave. The observatory started taking scientific data at the beginning of 2002 and is expected to release its first scientific results in 2003.

TAMA, VIRGO: These are, respectively, Japanese and European versions of LIGO. Due to design drawbacks, they are unlikely to be as sensitive as LIGO.

ALLEGRO, AURIGA: Unlike LIGO, which uses interferometry to detect gravitational waves, these experiments, and several others, are based upon a tuning-fork-like detector that vibrates when a gravitational wave of a certain frequency happens by. The detectors are less sensitive than TAMA and VIRGO.

LISA: A NASA vision for the ultimate gravitational-wave detector, the Laser Interferometer Space Antenna will be a formation of three satellites that act as an enormous interferometer. Unfortunately, the technical hurdles are formidable, but if LISA becomes reality, it would be an enormous boon to cosmologists trying to study gravitational waves from the early universe.

Neutrinos and WIMPs

Super-K: Though seriously damaged in late 2001, the Super-Kamiokande detector in Japan was the first to see convincing evidence that neutrinos have mass, and this announcement in 1998 was a watershed in neutrino physics. Basically an enormous cylinder of water studded with photodetectors,

Super-K detects telltale flashes of light as neutrinos interact with the water. Though Super-K will continue taking data, it will not be fully repaired for several years.

K2K: Two hundred and fifty kilometers away from Super-K, the KEK laboratory in Tsukuba, Japan, has been creating a beam of neutrinos that shoot toward the Super-K detector. Since 1999, the detector has been registering how many neutrinos it sees, and compares this quantity to the expected number; the difference is already revealing limits on neutrino masses. Though the experiment has suffered because of the Super-K damage, it will continue once Super-K is up to the task.

SNO: The Sudbury Neutrino Observatory released its first results in July 2001 and produced quite a reaction, as its researchers provided strong evidence that electron neutrinos from the sun were turning into muon and tau neutrinos as they stream toward Earth. This solved the solar neutrino paradox. Unlike Super-K, SNO is filled with heavy water, which makes it more sensitive to certain types of weak reactions. The results from SNO are likely to pin down many of the properties of neutrinos with great precision.

KamLAND: Using an old neutrino detector in the Kamioka mine in Japan, where Super-K is based, the KamLAND experiment is designed to detect antineutrinos that come from fission in the nuclear reactors that dot the Japanese and Korean countrysides. In December 2002, the KamLAND collaboration revealed its first results, which showed that antineutrinos oscillate just as neutrinos do. As the experiment gathers more data, physicists expect that the KamLAND team will dramatically improve scientists' knowledge about neutrino and antineutrino properties.

AMANDA, IceCube: An enormous neutrino detector made out of the Antarctic ice, the Antarctic Muon and Neutrino Detector Array has been measuring neutrinos and watching for WIMPs for the past few years. The equipment was upgraded in 1999 and 2000 and is still gathering and crunching data. AMANDA's planned successor, IceCube, has just begun to receive funding from the National Science Foundation.

Glossary

Cross-references are indicated with SMALL CAPS.

γ **(gamma):** A PHOTON.

Λ **(lambda):** The COSMOLOGICAL CONSTANT.

μ **(mu):** A MUON.

ν **(nu):** A NEUTRINO.

ν_e **(nu sub e):** An ELECTRON NEUTRINO.

ν_μ **(nu sub mu):** A MUON NEUTRINO.

ν_τ **(nu sub tau):** A TAU NEUTRINO.

π **(pi):** A PION.

τ **(tau):** A TAU PARTICLE.

Ω **(omega):** The density of "stuff" in the early universe—matter and energy. More technically, omega is the energy density of the universe, scaled by an appropriate factor to account for the expansion of the universe. (In this book, for the sake of clarity the scaling factor was ignored in discussions of omega.) Omega is related to the shape and the fate of the universe and is now thought to equal approximately 1.

Ω_b **(omega sub b):** The contribution of BARYONIC MATTER to the energy density of the universe. Scientists estimate that it equals about 0.05, or 5 percent. One-tenth of this is luminous matter; the rest is baryonic DARK MATTER.

Ω_m **(omega sub m):** The contribution of matter to the energy density of the universe. Scientists estimate that it equals about 0.35, or 35 percent. Most of this is EXOTIC DARK MATTER.

Ω_Λ **(omega sub lambda):** The contribution of the COSMOLOGICAL CONSTANT (or, more generally, DARK ENERGY) to the energy density of the universe. Scientists estimate that it equals about 0.65, or 65 percent.

Ω^- **(omega minus):** An OMEGA MINUS particle.

acoustic oscillations: The pressure waves that rattled around the early universe, as clumps of matter, under the attractive influence of gravity and the repulsive influence of radiation pressure, alternately compressed and expanded. These compressions and expansions are the source of the hot and cold spots in the cosmic microwave background.

anisotropy: The property of being different in different directions; the opposite of ISOTROPY. The discovery that the COSMIC MICROWAVE BACKGROUND is anisotropic was one of the key achievements of the COBE satellite.

antielectron: The antimatter twin of the ELECTRON; it is also known as a positron.

antimatter: An equal and opposite substance to matter; when matter comes into contact with antimatter, such as an ANTIELECTRON with an ELECTRON, the two annihilate, releasing energy.

baryon: A "heavy" particle, as compared to the lighter MESONS and LEPTONS, that is made up of three QUARKS.

baryonic matter: Matter made out of BARYONS—for example, PROTONS and NEUTRONS. All the matter encountered in everyday life, anything that is described in the periodic table of the elements, is baryonic matter.

beta decay: A nuclear process whereby a NEUTRON turns into a PROTON by ejecting an ELECTRON and an antineutrino. (This is beta-minus decay. There is also beta-plus decay, when a proton turns into a neutron by ejecting an ANTIELEC-TRON and a NEUTRINO.)

big bang: The beginning of the universe, as described by modern COSMOLOGY. Big bang theory has enormous empirical support, including the observations of the COSMIC MICROWAVE BACKGROUND and the understanding of the NUCLEOSYNTHESIS of light elements.

big crunch: The death of a universe in which it collapses upon itself, heats up, and disappears in a reverse BIG BANG.

big splat: The formation of our universe as described by ekpyrotic theory. See EKPYROTIC UNIVERSE.

blackbody spectrum: The spectrum of light emitted by a nonreflecting object at a certain temperature. The COSMIC MICROWAVE BACKGROUND, as predicted by theorists, has a blackbody spectrum.

black dwarf: A cooled-off WHITE DWARF star.

black hole: A dead massive star, collapsed to incredible density. A black hole is so dense that even light that ventures too close is unable to escape its grasp.

blueshift: The opposite of REDSHIFT.

B meson: A MESON made of a bottom QUARK with an up or down antiquark, or of a bottom antiquark with an up or down quark.

boson: An object with an integer SPIN $(0, \pm1, \pm2, \text{etc.})$. The force carriers in the STANDARD MODEL—the PHOTON, GLUON, W BOSONS, and Z BOSON—are all bosons. Bosons, unlike FERMIONS, can occupy the same quantum state at the same time.

brown dwarf: A failed star, one that was not massive enough to ignite. A possible candidate for the identity of MACHOS.

Casimir effect: The ability of the ZERO-POINT ENERGY, the particles constantly winking in and out of existence, to exert a force. Predicted by Dutch physicist Hendrik Casimir, the Casimir effect has been measured.

causally connected: The condition indicating that two objects have been able to exchange information; that is, light has had a chance to transit between the two. If two objects are not causally connected, they are unable to influence each other in any way.

Cepheid variable star: A type of star whose brightness fluctuates regularly. Cepheid variables are valuable because their brightness is related to how fast they fluctuate, so by measuring how long it takes for one to fluctuate, astronomers can determine how bright it is: it becomes a STANDARD CANDLE. Edwin Hubble used Cepheid variables to figure out distances to Andromeda and other GALAXIES.

CERN: The French acronym for the European Organization for Nuclear Research. CERN, based in Geneva, is one of the key particle physics laboratories in the world and is home to the LEP and LHC.

Chandrasekhar limit: A mass of 1.44 times the mass of our sun. A star exceeding the Chandrasekhar limit can no longer maintain its stability through ELECTRON pressure; it will go SUPERNOVA and collapse to a NEUTRON STAR, a QUARK STAR, or a BLACK HOLE.

COBE: The Cosmic Background Explorer, a NASA satellite that measured the COSMIC MICROWAVE BACKGROUND across the whole sky. Its major achievements were to show that this radiation was BLACKBODY and that it was ANISOTROPIC.

cold dark matter: DARK MATTER that is not moving particularly fast. The best current models of the formation of structure in the universe require that the lion's share of dark matter be cold.

color: An abstract property of QUARKS that is useful in understanding STRONG-FORCE interactions.

cosmic background radiation: See COSMIC MICROWAVE BACKGROUND.

cosmic microwave background (CMB): Also known as the cosmic background radiation, this is light that was released about 400,000 years after the BIG BANG. Stretched and attenuated over fourteen billion years, this radiation now appears as a nearly uniform hiss of microwaves coming from all areas of the sky. The cosmic microwave background carries a wealth of information about the early universe and is a vital tool for cosmology.

cosmic strings: Not to be confused with superstrings, cosmic strings are incredibly dense objects that are one possible source of TOPOLOGICAL DEFECTS. No evidence for cosmic strings has yet been found.

cosmological constant: Originally, a term that Einstein put into his equations of general RELATIVITY to ensure an unchanging universe. Presently, a candidate for the source of DARK ENERGY—probably caused by the ZERO-POINT ENERGY.

cosmology: The study of the universe as a whole, particularly its structure, its beginning, and its end.

curl: A mathematical term that quantifies the amount of "swirliness" in a field, such as the polarization of the COSMIC MICROWAVE BACKGROUND (CMB). ACOUSTIC OSCILLATIONS cannot create curl in the polarization of the CMB, but GRAVITATIONAL WAVES must, so scientists are hoping to find curl in the CMB in order to find the signature of gravitational waves in the early universe.

dark energy: The mysterious substance that appears to be opposing the force of gravity, inflating the universe at ever greater speeds. The prime candidates for the source of dark energy are QUINTESSENCE and the COSMOLOGICAL CONSTANT.

dark matter: Matter that is not luminous, that does not shine with light. Almost all the matter in the universe is dark.

Dirac neutrino: The standard concept of a NEUTRINO. A Dirac neutrino has an antimatter twin, unlike the MAJORANA NEUTRINO.

Doppler effect: A change in FREQUENCY due to the relative motion of the sender and receiver. With light waves, this effect causes REDSHIFT and BLUESHIFT.

ekpyrotic universe: An M-THEORY-based COSMOLOGY, proposed by Paul Steinhardt and colleagues, that has consequences similar to those of INFLATION but is subtly different. Most notably, it has the universe begin in a BIG SPLAT rather than a BIG BANG.

electron: The lightest and most common LEPTON.

electron neutrino: A variety of NEUTRINO that is involved mostly in reactions with ELECTRONS.

epicycle: In the GEOCENTRIC COSMOLOGY, the motion of planets in little circles inside their larger circular orbits; epicycles were necessary to explain how the planets moved in the sky.

exotic dark matter: DARK MATTER that is not baryonic. NEUTRINOS, which are not BARYONS, make up some fraction of this matter, but the bulk of it must be made from as-yet-undiscovered material, such as WIMPs.

false vacuum: A state in which the ZERO-POINT ENERGY is higher than it is today. Such a state of false vacuum would tend to inflate the universe quite rapidly and would decay quickly into the "true," modern-day vacuum.

fermion: An object with a half-integer SPIN ($\pm 1/2$, $\pm 3/2$, etc.). The fundamental constituents of matter in the STANDARD MODEL—QUARKS and LEPTONS—are fermions. Fermions, unlike BOSONS, cannot occupy the same quantum state at the same time.

fission: The splitting of an atomic nucleus. If the nucleus is heavier than iron, then the process tends to release energy. Atom bombs, which use uranium-235 or plutonium as the nuclei to be split, exploit this process.

flavor: A particle type. For example, there are six flavors of QUARKS: up, down, strange, charm, bottom, and top.

frequency: A measure of how long it takes for succeeding crests of a wavelike phenomenon to arrive; the higher the frequency, the more crests arriving per second. With light, the higher a PHOTON's frequency, the more energy it has.

fusion: The joining of two atomic nuclei. If the resulting nucleus is lighter than iron, then the process tends to release energy. Stars are powered by the fusion of hydrogen.

galaxy: A collection of stars, like the Milky Way. (In fact, the term *galaxy* comes from the Greek word for milk.)

geocentric cosmology: The idea that Earth is at the center of the solar system (and universe). This was the dominant COS-

MOLOGY in the West from ancient times until the late Renaissance. See PTOLEMAIC COSMOLOGY.

geodesic: The shortest path between two points on a smooth surface. On a plane, geodesics are straight lines. On a sphere, geodesics are great circles. Light always travels along geodesics in SPACETIME.

gluon: The elementary particle that carries the STRONG FORCE.

gravitational lens: An object that, because of its mass, bends light. A *strong* gravitational lens produces multiple images of a background object. A *weak* gravitational lens distorts background images but does not yield multiples.

gravitational wave: A "ripple" in SPACETIME that moves at the speed of light. Gravitational waves have a distinctive effect, squashing in one direction and stretching in another, that LIGO and other gravitational-wave laboratories hope to detect.

H_0: The present-day value of the HUBBLE CONSTANT. NASA estimates put it at about 72, but many astronomers (and cosmologists) tend to choose a slightly lower number, around 65. Precision measurements of the COSMIC MICROWAVE BACKGROUND will put the matter to rest within the next decade.

Heisenberg uncertainty principle: The law that states that certain linked properties of an object, such as position and MOMENTUM, cannot simultaneously be known with absolute precision. The principle is an unavoidable consequence of the mathematics of quantum theory.

heliocentric cosmology: A COSMOLOGY based upon the idea that the sun is at the center of the solar system. Copernicus started the shift from geocentrism to heliocentrism.

Higgs boson: A particle that imbues objects with mass. The Higgs should be found by the LHC.

hot dark matter: DARK MATTER whose energy is mostly tied up in motion rather than in mass. NEUTRINOS are one form of hot dark matter.

Hubble constant: A number that describes the rate of the HUBBLE EXPANSION, measured in units of km/sec/Mpc.

Hubble expansion: The expansion of SPACETIME, like a balloon. The effect of the Hubble expansion is to make GALAXIES appear to be receding from Earth, and the farther away the galaxies are, the faster they speed away. When the SUPERNOVA hunters discovered that the Hubble expansion was speeding up rather than slowing down, the third cosmological revolution began.

inflation: A theory proposed by Alan Guth to solve the horizon and flatness problems; it is a key part of big bang COSMOLOGY. The theory of inflation states that, from about 10^{-35} seconds to 10^{-32} seconds after the BIG BANG, the universe expanded extremely rapidly.

interferometer: An instrument that sends light along two (or more) paths and detects whether the beams interfere with or reinforce each other. Interferometry is a very sensitive tool for measuring changes in distance, and it is also useful in making sophisticated antennas and telescopes.

isotope: Two atoms are isotopes of each other if their nuclei have the same number of PROTONS but different numbers of NEUTRONS. For example, deuterium (one proton, one neutron) and tritium (one proton, two neutrons) are isotopes of hydrogen (one proton, zero neutrons).

isotropy: The property of being the same in every direction. Something that is isotropic looks the same in all directions of the sky; something that is anisotropic has asymmetries.

J/psi (J/ψ) particle: A MESON made up of a charm QUARK and a charm antiquark; so named because it was discovered (and named) by two groups at roughly the same time. One of the signs that scientists have created a QUARK-GLUON PLASMA is the suppression of J/psi particles.

jet quenching: The process whereby particle jets are reduced because they have to pass through a QUARK-GLUON PLASMA. Jet quenching in energetic collisions is one indicator to scientists that they have re-created the conditions shortly after the BIG BANG.

K meson: Also known as a kaon, it is a MESON made up of a strange QUARK with an up or down antiquark, or a strange antiquark with an up or down quark.

last scattering surface: The surface of the cloud of PLASMA that condensed during RECOMBINATION, so called because the PHOTONS that became the COSMIC MICROWAVE BACKGROUND took one last bounce, scattering off that plasma during recombination. When astronomers look at the CMB, they are actually forming an image of the last scattering surface.

LEP: The Large Electron-Positron collider, a now-defunct particle accelerator at CERN that was removed to make way for the LHC.

lepton: Unlike the heavier BARYONS and MESONS, leptons are fundamental, indivisible particles. There are six known leptons: the ELECTRON, MUON, TAU PARTICLE, and their associated NEUTRINOS.

LHC: The Large Hadron Collider, an expensive particle accelerator at CERN that is slated to begin operation by the end of the decade. It should discover the HIGGS BOSON.

light-year: A measure of distance; the distance that a beam of light travels in one year, or about six trillion miles. The nearest star to Earth is more than four light-years away. The nearest GALAXY is about a million light-years distant.

LIGO: The Laser Interferometer Gravitational-Wave Observatory, a GRAVITATIONAL-WAVE detector based in Washington and Louisiana.

LSP: The lightest supersymmetric partner. The LSP, if it exists, is a stable particle and might form the bulk of DARK MATTER.

MACHO: A massive compact halo object; a large lump of DARK MATTER (probably BARYONIC) in the dark matter halos that surround GALAXIES. The best candidates for MACHOs are dead or failed stars.

magnetic moment: A measure of how a particle twists in a magnetic field.

magnetic monopole: A hypothetical particle having only one magnetic pole. A magnet's positive (north) end cannot be separated from its negative (south) end; a bar magnet divided in two simply becomes two smaller, bipolar bar magnets. However, some theories imply that isolated norths or souths — magnetic monopoles — should have been created in the early universe.

Majorana neutrino: A NEUTRINO that is its own ANTIMATTER twin; in other words, there is no difference between a neutrino and an antineutrino in the Majorana formulation of the neutrino. The Majorana picture of the neutrino has some advantages, and scientists are trying to figure out whether neutrinos are Majorana or Dirac. If indeed the Majorana picture is correct, then there should exist a rare type of decay that has not yet been observed, called double beta decay. See DIRAC NEUTRINO.

many worlds hypothesis: The idea that our universe is one of many, many similar universes, each part of a MULTIVERSE.

MAP: The Microwave Anisotropy Probe, a microwave-background-sensing satellite launched in 2001. MAP is the successor to COBE and will, in turn, be succeeded by Planck.

meson: A middle-weight particle made up of a QUARK and an antiquark.

microlens: A gravitational lens produced by a small astronomical body that makes a background object brighten and then dim.

momentum: The "pushing power" of an object, typically a function of a particle's mass and velocity. (Light, though massless, has momentum too.)

MOND: Modified Newtonian Dynamics, a theory that alters Newtonian gravity slightly in an attempt to explain the motions of stars in GALAXIES without resorting to DARK MATTER.

M-theory: The eleven-dimensional unification of SUPERSTRING THEORY, which treats particles as "branes" rather than points. M-theory is the most promising candidate for providing a unified theory of all particles and forces.

multiverse: The all-encompassing structure of the MANY WORLDS HYPOTHESIS. It contains our universe, as well as numerous others.

muon: A LEPTON with a mass that falls between the mass of the ELECTRON and that of the TAU particle.

muon neutrino: A variety of NEUTRINO that is involved mostly in reactions with MUONS.

nebula: Originally, a "fuzziness" in the sky, and a term once applied to objects now known to be GALAXIES. Modern nomenclature is more specific: nebulae tend to be clouds of gas and not galaxies.

neutrino: A light LEPTON that interacts via the WEAK FORCE.

neutron: A neutral BARYON that is only slightly heavier than the PROTON. The neutron is unstable by itself; neutrons join with protons to make up atomic nuclei.

neutron star: A dead medium-size star, greater than the CHANDRASEKHAR LIMIT but less than the mass needed to form a QUARK STAR or BLACK HOLE.

nucleosynthesis: The creation of heavy nuclei out of PROTONS and NEUTRONS. The era of nucleosynthesis, which created most of the universe's helium, started a few seconds after the BIG BANG and lasted a few minutes.

omega, omega sub b, omega sub m, omega sub lambda: See pages 249–50.

omega minus: Not to be confused with the other omegas, the omega minus is a "strange" subatomic particle predicted by physicist Murray Gell-Mann and discovered shortly thereafter, providing strong support for the mathematical picture of subatomic particles that Gell-Mann was building.

"ordinary" matter: Also known as BARYONIC MATTER, the matter that people encounter in everyday life; stuff that is made up of atoms.

parallax: A method of determining distance that relies upon taking observations from two different perspectives.

parity: A mathematical term that is related to reflecting an object in a mirror—more accurately, in three mirrors, switching left for right, up for down, and front for back.

parsec: A measure of distance, about 3.26 LIGHT-YEARS. So named because it is a *parallax second;* an object that is one parsec away will appear to move one second (one-sixtieth of a degree) in the sky over the course of a year.

phase: A wave's phase is a description of whether a given point is experiencing a crest, a trough, or something in between. Two waves that are cresting at the same time at the same place are "in phase."

photon: A particle of light. Also the carrier of the electromagnetic force.

pion: Also known as a pi meson, this type of MESON has three varieties, made of different pairings of up and down QUARKS and their antiquarks.

plasma: A state of matter in which ELECTRONS are not bound to atomic nuclei.

polarization: A directionality to particles; light, for instance, can be polarized vertically, horizontally, or in other ways. This directionality can be detected by various means; light's polarization can be seen with the help of polarizing glasses.

positron: See ANTIELECTRON.

proton: A BARYON with a +1 charge. The proton is stable; protons join with NEUTRONS to make up atomic nuclei. A proton alone is also a hydrogen nucleus.

Ptolemaic cosmology: The most sophisticated and intricate GEOCENTRIC COSMOLOGY, which dominated Western thought until replaced by HELIOCENTRIC COSMOLOGY.

quark: A fundamental particle that makes up BARYONS and MESONS. There are six known flavors of quarks: up, down, charm, strange, bottom, and top.

quark-gluon plasma: A state of matter in which QUARKS and GLUONS roam free, rather than being confined in BARYONS and MESONS. It is thought that the quark-gluon plasma condensed into baryons at about a millionth of a second after the BIG BANG.

quark star: A hypothetical dead star. Also known as a strange star, a quark star is almost indistinguishable from a NEUTRON STAR. However, in a quark star the QUARKS and GLUONS are not confined in BARYONS, as they are in neutron stars.

quasar: A quasi-stellar object, a bright source of light that, like a star, seems quite small but is far too energetic to be an ordinary star. Modern theorists believe that a quasar is a massive, radiation-emitting BLACK HOLE at the center of a galaxy.

quintessence: 1) A hypothetical source of the mysterious antigravity force that suffuses the universe. Quintessence, which might be caused by an undetected particle, looks something like a time-varying COSMOLOGICAL CONSTANT. 2) The fifth element in ancient Greek COSMOLOGY that supplements the four others: earth, water, air, and fire.

recombination: A process that occurred 400,000 years after the BIG BANG, when the universe cooled enough so that ELECTRONS could combine with nuclei. Recombination released light from its cage of PLASMA, light that is now the COSMIC MICROWAVE BACKGROUND.

redshift: The shift of an object's light toward the red, less energetic, part of the spectrum as the object moves away from an observer. It is caused by the Doppler effect. Originally used to describe only light waves, the terms redshift and BLUESHIFT are now applied to other kinds of waves, such as GRAVITATIONAL WAVES.

reionization: A process that occurred hundreds of millions of years after the BIG BANG, when enough stars, GALAXIES, and QUASARS had formed to ionize the hydrogen "fog," ending the cosmic dark age.

relativity, theory of: A description of space and time devised by Einstein in the first two decades of the twentieth century. *Special* relativity deals with objects that are moving at a constant velocity, whereas *general* relativity deals with accelerating objects and gravity as well.

RHIC: The Relativistic Heavy Ion Collider at the Brookhaven National Laboratory in New York, a particle accelerator that has probably formed a QUARK-GLUON PLASMA.

Sachs-Wolfe effect: Gravitational kneading of PHOTONS caused by their journey into and out of a gravitational "dimple" that is changing size.

spacetime: The relativistic combination of space and time. Einstein's theory of RELATIVITY shows that space and time cannot be considered independently; they are functionally a single object. This object can curve and distort, and gravity can be considered a dimple in the "rubber sheet" fabric of spacetime.

sparticle: A supersymmetric particle, such as a neutralino or a squark. See SUPERSYMMETRY.

spectrum: 1) The colors of light produced when a light beam is passed through a prism; a light spectrum. 2) A term that mathematicians and scientists use to describe the breakdown of a mathematical object into its components. The bumpy plot of the angular size of features in the COSMIC MICROWAVE BACKGROUND is a spectrum of this sort, called a power spectrum.

spin: A quantum-mechanical property often likened to the spinning of a top. Whether an object has an integer $(0, \pm 1, \pm 2, \text{etc.})$ or half-integer $(\pm 1/2, \pm 3/2, \text{etc.})$ spin determines whether it is a BOSON or a FERMION.

standard candle: An object of known brightness. Standard candles, like CEPHEID VARIABLE STARS or TYPE IA SUPERNOVAE, are useful for gauging the distance to far-away objects.

standard model: A very successful mathematical model that describes the interactions of fundamental particles—the QUARKS, LEPTONS, and force carriers. Mathematically, the standard model describes the SYMMETRIES of an abstract seven-dimensional object.

standard ruler: An object of known size. Like STANDARD CANDLES, standard rulers are useful for measuring the distance to far-away objects and can also be used to measure the curvature of the universe.

strong force: The force, mediated by GLUONS, that binds QUARKS to each other, as well as binding the PROTONS and NEUTRONS in an atomic nucleus.

Sunyaev-Zel'dovich effect: A distortion in the COSMIC MICROWAVE BACKGROUND's spectrum caused by PHOTONS scattering off of hot ELECTRONS.

supernova: The violent death of a massive star. A supernova releases about 10^{51} ergs of energy, making it among the most energetic events in the universe.

superposition: The quantum-mechanical property of having two states at once. For example, an atom can be spin-up and spin-down at the same time, until something, like an observation, destroys that superposition—or in physicists' jargon, the "waveform collapses."

superstring theory: A set of ten-dimensional STANDARD-MODEL extensions that assume that fundamental particles, like ELECTRONS, are really strings rather than points. Superstring theories have been unified under the aegis of M-THEORY.

supersymmetry: An extension of the STANDARD MODEL that requires that each particle in the standard model have an undiscovered twin. Supersymmetry should be confirmed or falsified by the end of the LHC experiment.

symmetry: The notion of an object or a process remaining the same, even as it is changed in some way. For instance, a playing card is symmetrical because it looks exactly the same if rotated 180 degrees. The letter H is symmetrical because it looks the same when reflected in a mirror. The concept of symmetries is a fundamental idea that runs through modern physics.

symmetry group: A mathematical object that represents, in abstract form, the set of symmetries of a shape in space. The STANDARD MODEL, SUPERSYMMETRY, and many other important physical models are based upon the manipulation of symmetry groups.

tau neutrino: A variety of NEUTRINO that is involved mostly in reactions with TAU PARTICLES.

tau particle: A LEPTON similar to the ELECTRON and MUON, but considerably heavier than either.

tensor: A mathematical object that can be used to describe curvature. The equations of general RELATIVITY describe relationships between tensors.

topological defect: A defect in the smoothness of SPACETIME. Topological defects can be caused by many things, such as COSMIC STRINGS, and were once considered an alternative to INFLATION to explain the universe's structure. The spectrum of the COSMIC MICROWAVE BACKGROUND has recently ruled out topological defects as a major contributor to the early structure of the universe.

Tully-Fisher relation: A relationship between the speed at which a GALAXY rotates and its brightness. Discovered in the late 1970s, the Tully-Fisher relation turns galaxies into (not terribly accurate) STANDARD CANDLES.

type Ia supernova: A SUPERNOVA that occurs when an old, small star accretes mass from a companion star and exceeds the CHANDRASEKHAR LIMIT. These supernovae tend to have the same energy and are thus STANDARD CANDLES.

wavelength: The distance between succeeding crests in a wave. With light, the larger the wavelength of a PHOTON, the smaller its energy.

W boson: An elementary particle that carries the WEAK FORCE; there are two known varieties, W^+ and W^-, which carry a positive and a negative charge, respectively.

weak force: A force, carried by the W BOSONS and Z BOSON, that has the ability to change a particle from, say, an up QUARK to a down quark, or a NEUTRINO into an ELECTRON.

white dwarf: The last phase of a small star (our own sun will become a white dwarf). Larger stars become NEUTRON STARS, QUARK STARS, or BLACK HOLES.

WIMP: A weakly interacting massive particle; a prime candidate for the EXOTIC DARK MATTER in the universe. WIMPs might be LSPs.

W Virginis star: A type of CEPHEID VARIABLE, discovered by Walter Baade, that is dimmer than the classical Cepheid. Not knowing about W Virginis stars led to an error in Edwin Hubble's calculations.

Z: An astronomical (and nonlinear) measure of distance related to REDSHIFT. High-Z means large redshift.

Z boson: An uncharged elementary particle that carries the WEAK FORCE.

zero-point energy: The energy caused by the spontaneous creation and destruction of subatomic particles, even in the deepest vacuum. It is a prime suspect for the cause of the COSMOLOGICAL CONSTANT.

Select Bibliography

Books and Articles

Albrecht, Andreas, et al. "Early Universe Cosmology and Tests of Fundamental Physics: Report of the P4.8 Working Subgroup, Snowmass 2001." In arXiv.org e-Print archive (www.arxiv.org), hep-ph/0111080, 7 November 2001.

Anderson, C. D. "The Positive Electron." *Physical Review* 43 (1933): 491.

Arabadjis, J. S., et al. "Chandra Observations of the Lensing Cluster EMSS 1358+6245: Implications for Self-Interacting Dark Matter." In arXiv.org e-Print archive (www.arxiv.org), astro-ph/0109141, 19 February 2002.

Aristotle. *De Caelo*. J. L. Stocks, trans. Available at classics.mit.edu/Aristotle/heavens.1.i.html

———. *The Metaphysics*. John McMahon, trans. Amherst, N.Y.: Prometheus Books, 1991.

———. *Nichomachean Ethics*. H. Rackham, trans. Cambridge, Mass: Harvard University Press. 1934.

Augustine. *Confessions*. Henry Chadwick, trans. Oxford: Oxford University Press, 1991.

Bahcall, Neta, et al. "The Cosmic Triangle: Revealing the State of the Universe." *Science* 284 (1999): 1481.

Bautz, M. W., et al. "Chandra Observations and the Mass Distribution of EMSS 1358+6245: Toward Constraints on Properties of Dark Matter." In arXiv.org e-Print archive (www.arxiv.org), astro-ph/0202338, 18 February 2002.

Bania, T. M., et al. "The Cosmological Density of Baryons Form Observations of $^3He^+$ in the Milky Way." *Nature* 415 (2002): 54.

Bearden, I. G., et al. "Rapidity Dependence of Antiproton to Proton Ratios in Au+Au collisions at sqrt(s_{NN}) = 130 GeV." In arXiv.org e-Print archive (www.arxiv.org), nucl-ex/0106011, 13 June 2001.

Belli, P., et al. "WIMP Search by the DAMA Experiment at Gran Sasso." In arXiv.org e-Print archive (www.arxiv.org), hep-ph/0112018, 3 December 2001.

Biagioli, Mario. *Galileo Courtier.* Chicago: University of Chicago Press, 1994.

Blake, Chris, and Jasper Wall. "A Velocity Dipole in the Distribution of Radio Galaxies." *Nature* 416 (2002): 150.

Blandford, R. D. "Cosmological Applications of Gravitational Lensing." *Annual Review of Astronomical Astrophysics* 30 (1992): 311.

Blasi, P., et al. "Detecting WIMPs in the Microwave Sky." In arXiv.org e-Print archive (www.arxiv.org), astro-ph/0202049, 5 February 2002.

"Bush Finds Error in Fermilab Calculations." *The Onion,* 1 August 2001, 1.

Caldwell, Robert, and Paul Steinhardt. "Quintessence." Available at physicsweb.org/article/world/13/11/8

Charbonnel, Corinne. "A Baryometer is Back." *Nature* 415 (2002): 27.

Cho, Adrian. "Sign of Supersymmetry Fades Away." *Science* 294 (2001): 2449.

Christenson, J. H., et al. "Evidence for the 2π Decay of the K_2^0 Meson." *Physical Review Letters* 13 (1964): 138.

Cipra, Barry. "Shaping a Universe." *Science* 292 (2002): 2237.

Cowen, Ron. "A Dark Force in the Universe." *Science News,* 7 April 2001, 218.

Creighton, Jolien. "Listening for Ringing Black Holes." In arXiv.org e-Print archive (www.arxiv.org), gr-qc/9712044, 10 December 1997.

Dalal, Neal, et al. "Testing the Cosmic Coincidence Problem and the Nature of Dark Energy." *Physical Review Letters* 87 (2001): art. no. 141302.

Davidson, Keay. "Feud Overshadows Discovery: 2 Teams Detect Signs of First Galaxies Formed after Big Bang." *San Francisco Chronicle,* 4 August 2001, p. A2.

Eliade, Mircea. *From Primitives to Zen.* San Francisco: Harper and Row, 1977.

Ellis, John. "Why Does CP Violation Matter to the Universe?" *CERN Courier,* available at http://www.cerncourier.com/main/article/39/8/16

Ellis, George. "Maintaining the Standard." *Nature* 416 (2002): 132.

Erikson, Joel, et al. "Measuring the Speed of Sound of Quintessence." In arXiv.org e-Print archive (www.arxiv.org), astro-ph/0112438, 19 December 2001.

Farmelo, Graham, ed. *It Must Be Beautiful: Great Equations of Modern Science.* London: Granta Books, 2002.

Ferriera, Pedro. "The Quintessence of Cosmology." *CERN Courier,* available at www.cerncourier.com/main/article/39/5/11

Feynman, Richard P. *QED: The Strange Theory of Light and Matter.* Princeton: Princeton University Press, 1985.

———. "Space-Time Approach to Quantum Electrodynamics." *Physical Review* 75 (1949): 486.

Finkbeiner, Ann. "'Invisible' Astronomers Give Their All to the Sloan." *Science* 292 (2001): 1472.

Flambaum, V. V., and E. V. Shuryak. "Limits on Cosmological Variation of Strong Interaction and Quark Masses from Big Bang Nucleosynthesis, Cosmic, Laboratory and Oklo Data." In arXiv.org e-Print archive (www.arxiv.org), hep-ph/0201303, 18 February 2002.

Fox, Karen. *The Big Bang Theory.* New York: Wiley, 2002.

Freedman, Wendy, et al. "Final Results from the Hubble Space Telescope Key Project to Measure the Hubble Constant." *Astrophysical Journal* 53 (2001): 47.

Fritzsch, Harald. *Quarks: The Stuff of Matter.* New York: Basic Books, 1983.

Gamow, G. "The Origin of Elements and the Separation of Galaxies." *Physical Review Letters* 74 (1948): 505.

Gangui, Alejandro. "In Support of Inflation." *Science* 291 (2001): 837.

Gawiser, Eric, and Joseph Silk. "Extracting Primordial Density Fluctuations." *Science* 280 (1988): 1405.

Glanz, James. "Exploding Stars Point to a Universal Repulsive Force." *Science* 279 (1998): 651.

———. "Exploring Cosmic Darkness, Scientists See Signs of Dawn." *New York Times,* 4 August 2001, p. A1.

———. "Germans' Claim on Dark Matter Is Greeted with Skepticism." *New York Times,* 26 February 2002, p. F4.

———. "New Light on Fate of the Universe." *Science* 278 (1997): 799.

———. "No Backing Off From the Accelerating Universe." *Science* 282 (1998): 1249.

———. "A Second Hint of Symmetry Violation." *Science* 282 (1998): 2169.

Goldhaber, G., et al. "Timescale Stretch Parameterization of Type Ia Supernova B-Band Light Curves. In arXiv.org e-Print archive (www.arxiv.org), astro-ph/0104382, 24 April 2001.

Goldsmith, Donald. "Supernovae Offer a First Glimpse of the Universe's Fate." *Science* 276 (1997): 37.

Goldstein, J. H., et al. "Estimates of Cosmological Parameters Using the CMB Angular Power Spectrum of ACBAR." In arXiv.org e-Print archive (www.arxiv.org) astro-ph/0212517, 24 December 2002.

Graves, Robert. *The Greek Myths.* Vols. 1 and 2. New York: Viking, 1955.

Groom, D. E., et al. *Review of Particle Physics.* The European Physical Journal. C15, 1 (2000).

Guth, Alan. "An Eternity of Bubbles?" Available at www.pbs.org/wnet/hawking/mysteries/html/uns_guth_1.html

———. "Inflationary Universe: A Possible Solution to the Horizon and Flatness Problems." *Physical Review D* 23 (1981): 347.

Herodotus. *The Histories.* Aubrey de Selincourt, trans. London: Penguin Books, 1972.

Hewett, Paul, and Stephen Warren. "Microlensing Sheds Light on Dark Matter." *Science* 275 (1997): 626.

Iliev, Ilian, et al. "On the Direct Detectability of the Cosmic Dark Ages: 21-cm Emission from Minihalos." In arXiv.org e-Print archive (www.arxiv.org), astro-ph/0202410, 22 February 2002.

"In the Dark." *Science* 294 (2001): 1433.

Irion, Robert. "B-Meson Factories Make a 'Number From Hell.'" *Science* 291 (2001): 1471.

———. "LIGO's Mission of Gravity." *Science* 288 (2000): 5465.

———. "The Quest for Population III." *Science* 295 (2002): 66.

Kamionkowski, Marc, and Arthur Kosowsky. "The Cosmic Microwave Background and Particle Physics." In physics e-Print archive (www. arxiv.org), astro-ph/9904108, 9 April 1999.

Kane, Gordon. *Supersymmetry*. Cambridge, Mass.: Perseus, 2000.

Krauss, Lawrence. "Cosmology as Seen from Venice." In arXiv.org e-Print archive (www.arxiv.org), astro-ph/0106149, 8 June 2001.

Krauss, Lawrence, and Glenn Starkman. "Life, the Universe, and Nothing: Life and Death in an Ever-Expanding Universe." In arXiv.org e-Print archive (www.arxiv.org), astro-ph/9902189, 12 February 1999.

Krauss, Lawrence, and Michael Turner. "Geometry and Destiny." In arXiv.org e-Print archive (www.arxiv.org), astro-ph/9904020, 1 April 1999.

Kriss, G. A., et al. "Resolving the Structure of Ionized Helium in the Intergalactic Medium with the Far Ultraviolet Spectroscopic Explorer." *Science* 293 (2001): 1112.

Kuhn, Thomas. *The Structure of Scientific Revolutions*. Chicago: The University of Chicago Press, 1996.

Lahav, Ofer, et al. "The 2dF Galaxy Redshift Survey: The Amplitudes of Fluctuations in the 2dFGRS and the CMB, and Implications for Galaxy Biasing." In arXiv.org e-Print archive (www.arxiv.org), astro-ph/0112162, 7 December 2001.

Lee, T.-D., and C. N. Yang. "Question of Parity Conservation in Weak Interactions." *Physical Review* 105 (1957): 1671.

Lineweaver, Charles. "Cosmological Parameters." In arXiv.org e-Print archive (www.arxiv.org), astro-ph/0112381, 17 December 2001.

Lubin, Lori, and Allan Sandage. "The Tolman Surface Brightness Test for the Reality of the Expansion. I. Calibration of the Necessary Local Parameters." In arXiv.org e-Print archive (www.arxiv.org), astro-ph/0102213, 12 February 2001.

———. "The Tolman Surface Brightness Test for the Reality of the Expansion. II. The Effect of the Point-Spread Function and Galaxy Ellipticity on the Derived Photometric Parameters." In arXiv.org e-Print archive (www.arxiv.org), astro-ph/01012214, 12 February 2001.

———. "The Tolman Surface Brightness Test for the Reality of the Expansion. III. HST Profile and Surface Brightness Data for Early-Type

Galaxies in Three High-Redshift Clusters." In arXiv.org e-Print archive (www.arxiv.org), astro-ph/106563, 29 June 2001.

———. "The Tolman Surface Brightness Test for the Reality of the Expansion. IV. A Measurement of the Tolman Signal and the Luminosity Evolution of Early-Type Galaxies." In arXiv.org e-Print archive (www.arxiv.org), astro-ph/106566, 29 June 2001.

Manchester, William. *A World Lit Only by Fire.* Boston: Back Bay Books, 1993.

Miller, Christopher, et al. "Acoustic Oscillations in the Early Universe and Today." *Science* 292 (2001): 2302.

Miralda-Escude, Jordi. "Probing Matter at the Lowest Densities." *Science* 293 (2001): 1055.

Mohr, Joseph. "Probing the Distant Universe with the Sunyaev-Zel'-dovich Effect." Available at www.astro.uiuc.edu/~jmohr/Michelson/SZ_probe/

Morales, Angel. "Experimental Searches for Non-Baryonic Dark Matter: WIMP Direct Detection." In arXiv.org e-Print archive (www.arxiv.org), astro-ph/0112550, 27 December 2001.

Nagle, J. L., and T. Ullrich. "Heavy Ion Experiments at RHIC: The First Year." In arXiv.org e-Print archive (www.arxiv.org), nucl-ex/0103007, 15 March 2001.

Navick, X-F., et al. "Dark Matter Search in the EDELWEISS Experiment Using a 320 g Ionization-Heat Ge-Detector." Available at http://www-dapnia.cea.fr/Doc/Publications/Archives/dap-01-11.pdf

Normile, Dennis. "Weighing In on Neutrino Mass." *Science* 280 (1998): 1689.

Ovid. *Metamorphoses.* Rolfe Humphries, trans. Bloomington: Indiana University Press, 1955.

Pahre, Michael, et al. "A Tolman Surface Brightness Test for Universal Expansion and the Evolution of Elliptical Galaxies in Distant Clusters." *Astrophysical Journal* 456 (1996): L79.

Panek, Richard. *Seeing and Believing.* New York: Viking, 1998.

Parodi, B. R., et al. "Supernova Type IA Luminosities, Their Dependence on Second Parameters, and the Value of H_0." *Astrophysical Journal* 540 (2000): 634.

Penzias, A. A., and R. W. Wilson. "Measurement of the Flux Density of Cas A at 4080 Mc/s." *Astrophysical Journal* 142 (1965): 1149.

Percival, Will, et al. "The 2dF Galaxy Redshift Survey: The Power Spectrum and the Matter Content of the Universe." In arXiv.org e-Print archive (www.arxiv.org), astro-ph/0105252, 15 May 2001.

Perlmutter, S., et al. "Discovery of a Supernova Explosion at Half the Age of the Universe and Its Cosmological Implications. In arXiv.org e-Print archive (www.arxiv.org), astro-ph/9712212, 16 December 1997.

Plato, *Theatetus* and *Timaeus.* In *Plato in Twelve Volumes,* W. R. M. Lamb, Harold N. Fowler, Paul Shorey, and R. G. Bury, trans. Cambridge, Mass.: Harvard University Press, 1914–1935.

The Poetic Edda. Lee M. Hollander, trans. Austin: University of Texas Press, 1990.

Primack, Joel. "The Nature of Dark Matter." In arXiv.org e-Print archive (www.arxiv.org), astro-ph/0112255, 14 December 2001.

———. "Whatever Happened to Hot Dark Matter?" *Beam Line*, Fall 2001, 50.

Redondi, Pietro. *Galileo Heretic*. Princeton: Princeton University Press, 1987.

Reines, F., and C. L. Cowan. "Free Antineutrino Absorption Cross Section. I. Measurement of the Free Antineutrino Absorption Cross Section by Protons." *Physical Review* 113 (1959): 273.

Rubin, Vera. "Dark Matter in the Universe." *Scientific American*, March 1998, p. 106.

Rubin, Vera, and W. Kent Ford Jr. "Rotation of the Andromeda Nebula from a Spectroscopic Survey of Emission Regions." *Astrophysical Journal* 159 (1970): 379.

Schilling, Govert. "Deep-Space 'Filament' Shows Cosmic Fabric." *Science* 292 (2001): 1629.

———. "Signs of MACHOs in a Far-Off Galaxy." *Science* 287 (2000): 779.

Schwarzschild, Bertram. "Cosmic Microwave Observations Yield More Evidence of Primordial Inflation." *Physics Today*, July 2001, 16.

Seife, Charles. "BOOMERANG Returns With Surprising News." *Science* 288 (2000): 595.

———. "CERN Collider Glimpses SUpersymmetry—Maybe." *Science* 289 (2000): 227.

———. "CERN Stakes Claim on New State of Matter." *Science* 287 (2000): 949.

———. "Echoes of the Big Bang Put Theories in Tune." *Science* 292 (2001): 823.

———. "Elusive Particle Leaves Telltale Trace." *Science* 289 (2000): 527.

———. "Elusive Particles Yield Long-Held Secrets." *Science* 294 (2001): 987.

———. "Fly's Eye Spies Highs in Cosmic Rays' Demise." *Science* 288 (2000): 1147.

———. "Hubble Knows." *New Scientist*, 5 June 1999, 11.

———. "Masters of Infinity." *New Scientist*, 23 October 1999, 23.

———. "Microwave Telescope Data Ring True." *Science* 291 (2001): 414.

———. "Muon Experiment Challenges Reigning Model of Particles." *Science* 291 (2001): 958.

———. "Neutrino Oddity Sends News of the Weak." *Science* 294 (2001): 1433.

———. "New Collider Sees Hints of Quark-Gluon Plasma." *Science* 291 (2001): 573.

———. "No Turning Back." *New Scientist*, 31 October 1998, 21.

———. "Orbiting Observatories Tally Dark Matter." *Science* 293 (2001): 1970.

———. "Peering Backward to the Cosmos's Fiery Birth." *Science* 292 (2002): 2236.

———. "Polymorphous Particles Solve Solar Mystery." *Science* 292 (2001): 2227.

———. "Primordial Gas: Fog, Not Clouds." *Science* 276 (1997): 899.

———. "Rings Reveal a Supernova's Story." *Science* 287 (2000): 1580.

———. "'Tired Light' Hypothesis Gets Retired." *Science* 292 (2001): 2414.

———. "Troubled by Glitches, Tevatron Scrambles to Retain Its Edge." *Science* 295 (2002): 942.

———. *Zero: The Biography of a Dangerous Idea.* New York: Viking, 2000.

Sigg, Daniel. "Gravitational Waves." Available at www.ligowa.caltech.edu/P980007-00.pdf

Sturluson, Snorri. *The Prose Edda.* Jean Young, trans. Berkeley: University of California Press, 1954.

Verde, Licia, et al. "The 2dF Galaxy Redshift Survey: The Bias of Galaxies and the Density of the Universe." In arXiv.org e-Print archive (www.arxiv.org), astro-ph/0112161, 6 December 2001.

Wang, Limin, et al. "Cosmic Concordance and Quintessence." In arXiv.org e-Print archive (www.arxiv.org), astro-ph/9901388, 14 January 2000.

Watson, Andrew. "Pull of Gravity Reveals Unseen Galaxy Cluster." *Science* 293 (2001): 1234.

Webb, J. K., et al. "Further Evidence for Cosmological Evolution of the Fine Structure Constant." *Physical Review Letters* 87 (2001): art. no. 091301.

Weisberg, J. M., and J. H. Taylor. "Observations of Post-Newtonian Timing Effects in the Binary Pulsar PSR 1913+16." *Physical Review Letters* 52 (1984): 1348.

Wilson, Robert W. "The Cosmic Microwave Background Radiation." Nobel lecture, 8 December 1978.

Web Sites

The AMANDA project web site. amanda.berkeley.edu

The Brookhaven National Laboratory web site. www.bnl.gov

The Catholic Encyclopedia. www.newadvent.org/cathen/

The CERN web site. www.cern.ch

The Fermi National Accelerator Laboratory (Fermilab) web site. www.fnal.gov

The Gran Sasso National Laboratory web site. www.lngs.infn.it

Hu, Wayne. "The Physics of Microwave Background Anisotropies." background.uchicago.edu

The IceCube project home page. icecube.wisc.edu

The Kamioka Observatory web page. www-sk.icrr.u-tokyo.ac.jp/index.html

LIGO laboratory home page. www.ligo.caltech.edu

The MACHO Project home page. www.macho.mcmaster.ca

The MacTutor History of Mathematics archive. www-groups.dcs.stand.
    ac.uk/~history/

Nemiroff, Robert, and Jerry Bonnell. "Great Debates in Astronomy."
    antwrp.gsfc.nasa.gov/diamond_jubilee/debate.html

The Nobel e-Museum. www.nobel.se

The Stanford Linear Accelerator Center (SLAC) web site. www.slac.
    stanford.edu

The Sloan Digital Sky Survey web site. www.sdss.org

The Sudbury Neutrino Observatory home page. www.sno.phy.queensu.ca

The 2dF Galaxy Redshift Survey web site. msowww.anu.edu.au/2dFGRS/

The UK Dark Matter Collaboration web site. hepwww.rl.ac.uk/ukdmc/
    ukdmc.html

Weisstein, Eric. "Treasure Troves of Science." www.treasure-troves.com

White, Martin. Online cosmology papers. astron.berkeley.edu/~mwhite/
    htmlpapers.html

Acknowledgments

A lot of people helped me write this book; it's not possible for me to name them all. Over the past few years I have interviewed dozens of physicists, cosmologists, and astronomers who took the time to explain the nuances of their work to a journalist. I thank them for their enthusiasm and their patience. They are the reason that I wrote *Alpha and Omega* in the first place. (Of course, they bear no responsibility for any errors in this work — any mistakes are mine alone.)

I would also like to thank my editor, Wendy Wolf; my copyeditor, Don Homolka; and my agents, John Brockman and Katinka Matson. Last but not least, my parents, once again, have been an unwavering source of support (and constructive criticism) even through the most difficult times of their lives. Thank you for everything.

Index